MEMORIZING PHARMACOLOGY

*A Relaxed Approach
Audiobook Script*

Tony Guerra, Pharm.D.
2017

Cpyright © 2017 by Tony Guerra, Pharm.D.

All rights reserved. No part of this book may be reproduced or transmitted in any form or by any means without written permission of the author.

Memorizing Pharmacology:
A Relaxed Approach Audiobook Script
By Tony Guerra, Pharm.D.
First Edition
ISBN: 978-1-387-05174-8

To Mindy,

Brielle, Rianne, and Teagan

Table of Contents

AUTHOR'S NOTE xiii
   The story of pharmacology class.................................. xiii
   The Curse of Knowledge... xvii
   Taking pharmacology twice, sort of. xviii
   Creating a primer xix
   Medication reconciliation (Med Rec)........................... xxi
   Final note xxi
ACKNOWLEDGEMENTS.. xxiii
INTRODUCTION... xxv
   Setting the scene xxv
   Generic names versus brand names xxvi
   Three types of drug names... xxviii
   Brand name rule of thumb: Can indicate function xxix
   Generic name rule of thumb: Can indicate class.......... xxx
   Prefixes, suffixes, and infixes... xxxi
   Pronunciation and organic chemistry word parts...... xxxii
   Homophones... xxxv
   Mnemonics in pharmacology xxxvi
   3 by 5 notecards xxxvii
   Comprehensive drug list discussion xxxviii
   OTC SCAVENGER HUNT xxxix
CHAPTER 1: GASTROINTESTINAL................................ 1
   Gastrointestinal Section I. Peptic ulcer disease 1

i

Histamine 2 receptor antagonists (H 2 R As) 3
Proton pump inhibitors (PPIs) 4

Gastrointestinal Section II. Diarrhea, constipation, and emesis 5

Two antidiarrheals 6
Two medications for constipation 7
Two antiemetics 8

Gastrointestinal section III. Gastrointestinal autoimmune disorders 8

Chapter 1 Quiz: Gastrointestinal, Drug stem and Drug classification practice 9

GI: Memorizing the chapter 10
"G" Gastrointestinal 11

CHAPTER 2: MUSCULOSKELETAL 13
Musculoskeletal section I. NSAIDs and pain 13

Three OTC NSAID Analgesics 14
A non-narcotic OTC Analgesic 15
OTC Migraine – An NSAID with a Non-narcotic analgesic and vasoconstrictor 16
A Prescription only NSAID Analgesic 16
A COX-2 inhibitor Prescription only NSAID analgesic
.... 16

Musculoskeletal section II. Opioids and narcotics 17

Four Schedule II Opioid analgesics 17
Opioid analgesic – Schedule III 19
Mixed-opioid receptor analgesic – Schedule IV 19
AN Opioid antagonist 20

Musculoskeletal section III. Headaches and migraines ... 20

We'll pair the triptans to better remember them. 21

Musculoskeletal section IV. DMARDs and rheumatoid arthritis 22

Table of Contents

The Non-biologic DMARD 22
There are two biologic DMARDs we'll talk about. 22

Musculoskeletal section V. Osteoporosis 23

Two Bisphosphonates for osteoporosis 23

Musculoskeletal section VI. Two Muscle relaxants 24

Musculoskeletal section VII. Gout 24

Two Uric acid reducers 25

Chapter 2 Quiz: Musculoskeletal, Drug stem and Drug classification practice 25

Musculoskeletal: Memorizing the chapter 27

"M" Musculoskeletal 27

CHAPTER 3: RESPIRATORY 29

RESPIRATORY SECTION I. Antihistamines and decongestants 29

An OTC 1st-generation Antihistamine 30
Two OTC 2nd-generation Antihistamines 31
A Combination 2nd-generation OTC Antihistamine and Decongestant 31
Three Decongestants that don't require a prescription 31

RESPIRATORY SECTION II. Allergic rhinitis steroid, antitussives and mucolytics 32

OTC Allergic rhinitis steroid nasal spray 33
OTC Mucolytic with Antitussive 33
Prescription-only Mucolytic with narcotic Antitussive 34

RESPIRATORY SECTION III. Asthma 34

Two Oral steroids 35
Two Inhaled steroid with long-acting Beta$_2$-receptor agonist combinations 36
An Inhaled steroid 37

A Beta 2 receptor agonist short-acting 37
Beta₂ receptor agonist with short-acting Anticholinergic
.... 38
Asthma and Chronic Obstructive Pulmonary Disease
(C-O-P-D) medication – long-acting anticholinergic 38
Asthma medication - Leukotriene receptor antagonist
.... 39
Asthma – Anti-I-g-E antibody 39
RESPIRATORY SECTION IV. Anaphylaxis 40
Chapter 3 Respiratory, Drug stem and Drug classification
practice 40
Respiratory: Memorizing the chapter 42
"R" Respiratory 42
CHAPTER 4: IMMUNE 45
IMMUNE SECTION I. OTC Antimicrobials 45
Second, let's look at an OTC Antifungal cream 46
Third, let's look at two types of antiviral, one antiviral
for prophylaxis (preventing infection) and one antiviral
to treat an active infection. 46
OTC Antiviral (Acute infection) 47
IMMUNE SECTION II. Antibiotics affecting cell walls .. 47
Antibiotics: Penicillins 49
A Penicillin class antibiotic with a Beta-lactamase
inhibitor 50
1st generation Cephalosporin antibiotic 50
3rd generation Cephalosporin antibiotic 51
4th generation cephalosporin antibiotic 51
Glycopeptide antibiotic 52
IMMUNE SECTION III. Antibiotics – Protein synthesis
inhibitors – bacteriostatic 52
Three Macrolide antibiotics 54
A Lincosamide antibiotic 55
An Oxazolidinone antibiotic 55

Table of Contents

IMMUNE SECTION IV. Antibiotics – Protein synthesis inhibitors – bactericidal 55

    Two Aminoglycoside antibiotics 56

IMMUNE SECTION V. Antibiotics for urinary tract infections (U-T-Is) and peptic ulcer disease (P-U-D) 56

    Dihydrofolate reductase inhibitor antibiotics 56
    Dihydrofolate reductase inhibitor combination antibiotic 57
    Let's look at two Fluoroquinolone antibiotics 58
    Nitroimidazole antiprotozoal 59
    IMMUNE SECTION VI. Anti-tuberculosis agents 59

IMMUNE SECTION VII. Antifungals 60

IMMUNE SECTION VIII. Antivirals – Non-HIV 62

    Two Oral Influenza A and B antivirals 62
    Two Herpes simplex virus (HSV) & Varicella-Zoster virus (VZV) antivirals 63
    Respiratory syncytial virus (RSV) 64

IMMUNE SECTION IX. Antivirals – HIV 64

    1) Fusion inhibitor 65
    2) Cellular chemokine receptor (CCR5) antagonist 65
    3) One Non-nucleoside reverse transcriptase inhibitor (N-N-R-T-I) with 2 Nucleoside forward slash Nucleotide reverse transcriptase inhibitors (N-R-T-Is) 65
    4) Integrase strand transfer inhibitor 66
    5) Protease inhibitor 66

Chapter 4 Quiz: Immune, Drug stem and drug classification practice 67

Immune: Memorizing the chapter 69

    "I" Immune 69

Memorizing Pharmacology

CHAPTER 5: NEURO 73

  NEURO SECTION I. OTC Local anesthetics and antivertigo 73

    Local anesthetic – Ester type 74
    Local anesthetic – Amide type 74
    Antivertigo 74

  NEURO SECTION II. Sedative-hypnotics (sleeping pills) 75

    OTC - Non-narcotic analgesic with Sedative-hypnotic 77
    Two Benzodiazepine-like sedative hypnotics 77
    A melatonin receptor agonist sedative hypnotic 78

  NEURO SECTION III. Antidepressants 79

    Selective serotonin reuptake inhibitors (SSRIs) 80
    Two Serotonin-norepinephrine reuptake inhibitors (SNRIs) 82
    Tricyclic antidepressant (TCA) 83
    Monoamine oxidase inhibitor (MAOI) antidepressant 83

  NEURO SECTION IV. Smoking cessation 84

  NEURO SECTION V. Benzodiazepines 85

  NEURO SECTION VI. Attention Deficit Hyperactivity Disorder (A-D-H-D) medications 87

    Two Stimulant DEA Schedule II ADHD medications 87
    Non-stimulant – non-scheduled ADHD medication ... 88

  NEURO SECTION VII. Bipolar disorder 89

    Simple salt – mood stabilizer 89

  NEURO SECTION VIII. Schizophrenia 90

    First generation antipsychotic (F-G-A) low potency ... 90
    First generation antipsychotic (F-G-A) high potency .. 91
    Second-generation antipsychotics (S-G-As) (Atypical antipsychotics) 91

Table of Contents

NEURO SECTION IX. Antiepileptics 92
   Three traditional antiepileptics 92
   Two Newer antiepileptics 93
NEURO SECTION X. Parkinson's, Alzheimer's and motion sickness 94
   Two Parkinson's medications 94
   Two Alzheimer's Medications 95
   A Motion sickness medication 96
Chapter 5 Quiz: Neuro, Drug stem and drug classification practice 96

Neuro: Memorizing the chapter 98
   "N" Neuro 98

CHAPTER 6: CARDIO 101
CARDIO SECTION I. OTC Antihyperlipidemics and antiplatelet 101
   Two OTC Antihyperlipidemics 101
   OTC Antiplatelet 102

CARDIO SECTION II. Diuretics 102
   Osmotic diuretic 103
   Loop diuretic 103
   Thiazide diuretic 104
   Potassium sparing with thiazide diuretic combination 104
   Potassium sparing diuretic 105
   Electrolyte replenishment 105
CARDIO SECTION III. Understanding the alphas and betas 106
   Alpha-1 antagonist 108
   Alpha-2 agonist 109
   Beta blockers – 1st-generation – non-beta-selective ... 109
   Beta blockers – 2nd-generation – beta-selective 110

Memorizing Pharmacology

    Beta blockers – 3rd-generation – non-beta-selective, vasodilating 111

CARDIO SECTION IV. The renin- angiotensin-aldosterone -system drugs 111

    TWO Angiotensin converting enzyme inhibitors (ACEIs) 113
    Three Angiotensin II receptor blockers (ARBs) 113

CARDIO SECTION V. Calcium channel blockers (CCBs) 114

    Two Non-dihydropyridines 115
    Two Dihydropyridines 115

CARDIO SECTION VI. Vasodilator 116

CARDIO SECTION VII. Antihyperlipidemics 116

    Two H-M-G-Co-A reductase inhibitors 117
    Fibric acid derivative 117

CARDIO SECTION VIII. Anticoagulants and antiplatelets 118

    Let's look at the four Anticoagulants Enoxaparin, Heparin, Coumadin, and Dabigatran 118
    Antiplatelet 120

CARDIO SECTION IX. Cardiac glycoside and anticholinergic 120

    Cardiac glycoside 120
    Anticholinergic injection 121

Chapter 6 Quiz: Cardio, Drug stem and drug classification practice 122

Cardio: Memorizing the Chapter 124

    "C" Cardio 124

Cardiode to Joy 125

Table of Contents

CHAPTER 7: ENDOCRINE AND MISCELLANEOUS 127

ENDOCRINE and MISCELLANEOUS SECTION I. OTC insulin and emergency contraception 127

- Regular insulin and N-P-H insulin............................ 127
- The Insulin Orange 127
- Over the counter emergency contraception 128
- Over the counter emergency contraception 129

ENDOCRINE AND MISCELLANEOUS SECTION II. prescription only diabetes medications 130

- Oral anti-diabetics - Biguanide..................................... 131
- Oral anti-diabetics– DPP-4 inhibitor 131
- Oral anti-diabetics – 2nd-generation Sulfonylureas.. 132
- Hypoglycemia treatment 132
- Two Prescription-only Insulins, insulin lispro, ultra-short acting and insulin glargine, long acting............ 133

ENDOCRINE AND MISCELLANEOUS SECTION III. Thyroid hormones... 134

- Hypothyroidism... 134
- Hyperthyroidism.. 134

ENDOCRINE AND MISCELLANEOUS SECTION IV. Hormones and contraception 135

- Testosterone replacement 135
- Contraception – 2 Combined oral contraceptives 136
- Contraception – Patch... 136
- Contraception – Ring.. 137

ENDOCRINE AND MISCELLANEOUS SECTION V. Overactive bladder, urinary retention, erectile dysfunction, benign prostatic hyperplasia 137

- Three Overactive bladder medications....................... 138
- A Urinary retention Medication................................... 138
- Erectile dysfunction (Two P-D-E-5 inhibitors)........... 139

Two alpha blockers for benign prostatic hyperplasia (BPH) 140
A Pair of 5-alpha-reductase inhibitors for benign prostatic hyperplasia (B-P-H) 140
Chapter 7 Quiz: Endocrine, Drug stem and drug classification practice 141
Endocrine and Miscellaneous: Memorizing the chapter 142
 "E" Endocrine 142

CHAPTER 8: MEMORIZING THE BOOK 145

Memorizing the book Part I 145
 "G" Gastrointestinal 145
 "M" Musculoskeletal 146
 "R" Respiratory 146
 "I" Immune 147
 "N" Neuro 149
 "C" Cardio 150
 "E" Endocrine 150

Memorizing the book Part II 152

CHAPTER 9: END OF SEMESTER 159

Introduction – Basic principles 159

Chapter 1 – Gastrointestinal 163

Chapter 2 – Musculoskeletal 165

Chapter 3 – Respiratory 167

Chapter 4 – Immune 169

Chapter 5 – Neuro 172

Chapter 6 – Cardio 175

Chapter 7 – Endocrine / Misc. 179

Table of Contents

EPILOGUE 183
  Why memorization matters 183
  There will be no problems 183
GENERIC AND BRAND NAME INDEX 189

Memorizing Pharmacology: A Relaxed Approach

Written by Tony Guerra, Pharm D

Narrated by James Gillies

AUTHOR'S NOTE

THE STORY OF PHARMACOLOGY CLASS

As an instructor, I didn't appreciate how tough it was to be a working parent and pharmacology student until I had my triplet daughters. When the girls originally came home, they had trouble coordinating the suck, swallow, breathe that goes with feeding from a bottle, well, three bottles. We fed them for 90 minutes and they slept for 90 minutes – round the clock. They got older, but the crushing emotional and physical exhaustion has continued. My students also have jobs and families. I wanted a way for them to study pharmacology while attending to these types of responsibilities.

Tonight, as I lay quietly in bed next to my daughter, who will not sleep otherwise, **I recited in my head, from memory, the 200 drugs, generic name, brand name and drug classification, in this book, in order.** Previously during these times, thoughts of "What do I have to do for tomorrow?" ran through my head. Instead, I could relax, stay unhurriedly by my daughter, and remain a committed parent. With this book's techniques, I could study pharmacology in the dark – eyes closed. I knew I could help my overworked students.

After she fell asleep, I got up and started writing this introduction, but before I could congratulate myself, my other daughter came downstairs insisting that she will never sleep in her bed again. So, here I am, on my couch, half-watching *Better Call Saul*, comforting another daughter between coughing fits, and typing out this introduction, knowing I might see the sunrise before I get to sleep.

Earlier this fall a student emailed me, offering to pay me to read and record his "Top 200 Drugs" list. I was too busy midsemester, but I did want to create a road map for him and other students that shared his need. Each fall and spring, many health professional students try to learn the "Top 200 Drugs," but get little help in memorization techniques.

This book *will* help, regardless of which Top 200 Drugs your professor assigns. It will also help you answer the question: where do I start studying pharmacology for my board exams? Maybe you're studying for the N-CLEX-R-N (National Council Licensure Examination – Registered Nurse) or NAB-PLEX (National Association of Boards of Pharmacy Licensure Examination) or another exam. This book will provide a framework for your knowledge to wrap around, much as a garden lattice might hold up vines.

What does Top 200 Drugs mean? At one time, there was a website that listed The Top 200 Drugs in the United States by 1) Number of prescriptions written and 2) Ranked by money spent on each drug. This created two different lists, but their value was clear. **Well-prepared students remember the most frequently prescribed drugs.** The 80/20 rule, or Pareto's principle, predicts that 20% of the medications will represent 80% of those prescribed.

Brutal rote memorization, however, is a poor strategy for managing information overload. Memorizing a brand name, generic name, medication class, therapeutic use, and one adverse effect for 200 drugs represents one-thousand pieces of information. Each new item adds 200 memorization points. Students often ask why they should memorize drugs if they can Google them or look them up in a *Davis Drug Guide for Nurses*. The answer: a properly sorted and

AUTHOR'S NOTE

memorized list of 200 drugs provides a structure on which to build your preparation for pharmacology class, the board exams, and clinical practice. But, how did *I* memorize so many drugs? Follow this thought experiment.

Imagine, instead of playing the part of student, you are the instructor. The class has 200 students and there are exactly 200 seats in the classroom. How do you remember all of their names? How do you know who is absent when you see empty chairs? You could ask the students to sit alphabetically by first name or by last name, but this is college, not grade school. A more organic approach would take time to understand why groups formed as they did.

Students sit in the same seats weekly. As you talk and get to know them, you find out what brings them together. They may have the same undergraduate major, hometown, dorm, previous class, and so forth. There are the front row groups that always asks questions or back row group that asks no questions. Groups come from the same hospital floor or work on a semester project. If someone's absent, you know what group he or she is missing from.

As you learn the drugs in this book, you'll see similar groupings. For example, you'll see thirteen gastrointestinal drugs, five SSRI antidepressants, four benzodiazepines, and three angiotensin-converting enzyme inhibitors. When you bring medicines back from memory, you'll remember if one is missing.

This book provides that specific framework to memorize the most important drugs in a logical order. A good analogy is that learning to drive a car requires essentially the same foundational instruction. However, once you learn, you can drive anywhere (or anything) you want.

We have a problem, however, because pharmacology instructors and students speak different languages. This causes instructors to get student comments like "He can't teach," "I didn't learn anything in his class," and "C's get degrees, I guess." Professors who can't to students frustrate those who want to learn. I'm especially empathetic to parents and students with full-time jobs and international students trying to pick up English *and* the language of pharmacology. I am a parent of three daughters and English is not my first language. Let me show you this disconnect between a student's and a teacher's memorization metaphors.

Pharmacology breaks down to "pharmaco" or "drug" and "logy" which is "study of" with a connecting "o" to make "study of drugs." There is another concept called *pharmacokinetics*. "Pharmaco" means "drug" and "kinetics" means movement, like a kinesiology major is someone who studies – "logy" – body movement – "kinesio."

When, as a pharmacology instructor, I look at the word *pharmacokinetics*, I think of four major principles: *absorption*, *distribution*, *metabolism*, and *excretion*. I remember the A-D-M-E mnemonic, pronounced: "add-me." I remember each principle as matching a certain organ or tissue.

- A: Absorption happens in the *small intestine's* villi.
- D: Distribution occurs via the *blood*.
- M: Metabolism happens in the *liver*.
- E: Excretion occurs through the *kidneys*.

With my anatomy and physiology background, I memorized these facts through a metaphor of connected tunnels. Drugs travel from the mouth to the small intestine to be *absorbed*. Blood vessels *distribute* the drug to the liver, which *metabolizes* it. Eventually, the body *excretes* the drug through the kidney. Finally, the ureters carry the drug to the bladder and the urethra. That picture runs clearly through my brain.

I assumed my students, some veterans of anatomy and physiology, learned the same way. That wasn't true. When asked to create a writing prompt of how they memorized the same idea, my students used a different set of metaphors.

AUTHOR'S NOTE

One equated *absorption* with soaking up material in a classroom, *distributing* knowledge to short- and long-term memory, *metabolizing* or breaking down information into manageable topics, and *excretion* to eliminating extra pieces of information.

Another student condensed a semester's worth of developmental psychology into a sentence. She associated the increased surface area of her pregnant belly with the intestine's *absorption* and *distribution* with the bloodline her child inherits. She credited life changes and possibly future drinking after pregnancy, to remember the liver's *metabolism*. *Excretion* matched to her child leaving home after high school.

After reading that story, can you tell me what *pharmacology* and *pharmacokinetics* mean? Whose story did you use, the instructor's or the student's? I don't think either is better. I believe it's helpful to have two points of view.

THE CURSE OF KNOWLEDGE

The expert-novice divide has a name – "The Curse of Knowledge." Experts understand the whole, and often can't explain it step-by-step. Pharmacology works identically.

Whatever pharmacology text you're using has an author trying to tell the same story from a different expert point of view to novice audiences. I've found pharmacology books specifically titled for allied health, athletic training, audiologists, dental hygienists, E-M-S providers, health professions, massage therapists, medical assistants, medical office workers, nurses, paramedics, pharmacists, pharmacy technicians, physical therapists, doctors, rehabilitation professionals, respiratory care therapists, surgical technologists, veterinary technicians and veterinarians. I stopped counting at twenty. Even as someone writes a textbook, the base of knowledge expands and changes. The better approach starts with a general primer with a small amount of information, only 200 drugs. *Then*, a student can take what he or she specifically needs from discipline-specific pharmacology textbooks.

In Britain, there is a book that unifies the medications the health professions employ in practice – the British National Formulary or B-N-F. I didn't know about the British National Formulary before writing this book, but having one central volume that multiple professions agree on makes it much easier to collaborate inter-professionally.

Learning pharmacology might be easier if multiple health professions agree on a central drug primer.

I've read many discipline-specific books and taught in many fields, but you get the point. Pharmacology is an ever-expanding subject too big to distill. It's best learned by building a central base and building up from there. I'd recommend that you find out how your classmates learn, but in a big lecture hall, you often don't talk to them. *This* book picks up the shortcuts and methods successful students employ to do well in pharmacology.

TAKING PHARMACOLOGY TWICE, SORT OF.

Most pharmacology textbooks have the same basic format. The early chapters cover the interactions of the drugs with the body, *pharmacodynamics*, and the body on the drug, *pharmacokinetics*, and introduces some basic vocabulary. The author divides chapters by pathophysiologic condition and medications follow. For example, antibiotic medications follow gastrointestinal medicines.

However, many sections rely on understanding material from a *future* section. The instructor has a full semester's overview. Student's don't. Soon, the instructor talks over

AUTHOR'S NOTE

the student's head. Here's an example: What medications treat an ulcer? Why use these medications?

A student can memorize the three medications, **omeprazole (brand name Prilosec), amoxicillin (brand name Amoxil),** and **clarithromycin (brand name Biaxin)** that comprise an ulcer treatment regimen. However, to move up Bloom's taxonomy from knowledge (memorizing "what") to comprehension (understanding "why") requires significant added information.

- **omeprazole,** a proton pump inhibitor, reduces stomach acid and gains an advantage over the antacid **calcium carbonate** and histamine two receptor blocker **ranitidine** because of its extended half-life. Half-life measures the time it takes for a drug in the body to reduce by half.
- **amoxicillin** kills the causative agent in most ulcers, *Helicobacter pylori*, a helicopter-shaped bacterium that's sensitive to penicillin class antibiotics.
- **clarithromycin** reduces the incidence of resistance that can happen with a single broad-spectrum antibiotic like **amoxicillin.**

Instructors teach **omeprazole** in the earlier gastrointestinal section. *Weeks later*, students learn how **amoxicillin** and **clarithromycin** work in the antimicrobial section of the course. This second major point is critical – **To learn pharmacology requires that a student has already taken pharmacology.** This seems ridiculous, but let's use an apt metaphor, a trip to another country.

CREATING A PRIMER

Imagine you'll start college in the fall and have decided to take a four-credit Spanish class. You could hope the teacher slows down to whatever speed you need, or you could study ahead of time.

Memorizing Pharmacology

Let's pretend you decide to travel to the Peruvian mountains – my father's homeland. Although some people speak English, you ask them to speak to you in Spanish so they can help you learn Spanish for your class. Your smartphone doesn't get any signal in the Andes Mountains, so you use pen and paper to write your notes in a small journal to pick up words along the way. At first, you point to objects, but, gradually, you can start making sentences. You make notes of words that are especially tricky.

For example, the Spanish word embarazada *looks like* the English word embarrassed, but it *means* pregnant. If you make this mistake, your brain will remember the story of that mistake, and you won't make it again. You compare notes with other students and share stories each night. In the end, you have a primer, an introductory book in Spanish that will put you far ahead of your class as you enter college.

The word "primer" has different meanings in various contexts. In auto painting and cosmetics, a primer allows for better adhesion for a secondary product's application. A primer is also a basic language-reading textbook. You can combine these definitions to create a useful analogy. **A pharmacology primer can provide a home base for students to expand a basic framework of vocabulary before they use PowerPoints (sentences) and textbooks (full paragraphs).**

This book is your journal. In the same way you take medical terminology before you take anatomy and physiology, you want to master some of the terms before you start pharm class. The following seven chapters include conversations with students about creating meaning from words in the foreign language of pharmacology. I'll talk through how and why I've grouped them as I did. Once you've completed this book, you should be able to use these drugs conversationally.

MEDICATION RECONCILIATION (MED REC)

While this book's focus is to prepare students, I believe it can help patients and caregivers provide accurate medication histories and records – as students of their own conditions. Imagine if every patient knew not 20 drugs, but 200. What if they understood where their therapy fit into the greater body of medications?

Medical reconciliation or "med rec" is the process of assembling a correct list of patients' medications under what might be some stressful circumstances. When an ambulance comes, you may forget to take everything with you. Memorizing medications in a logical order ensures the list will always be with you and you can immediately tell EMS responders what medications you or the patient you care for is on.

FINAL NOTE

If you invent a great mnemonic and want to share it for a future edition, feel free to email me:

aaguerra@dmacc.edu

I'm always looking for better ways to teach pharm. If you see a student struggling with the language of pharmacology who doesn't have this book or a patient who wants to better understand his or her many prescriptions, please recommend it to them.

There was a pharmacology class that I got a standing ovation, just like in the 1972 movie *The Paper Chase*. I hope what you get out of this book is worth your standing up.

Tony Guerra
December 28th, 2015
2:41 AM – Still dark outside

ACKNOWLEDGEMENTS

I could never have completed this book without the support and love of my wife Mindy, or the continued warmth and lessons I receive from my daughters Brielle, Rianne, and Teagan.

INTRODUCTION

SETTING THE SCENE

Imagine we'll meet at a large six foot by six foot wooden table in the college library. We are side-by-side with textbooks open and laptops or electronic tablets sitting diagonal to us.

Some topics we'll talk about you'll just need to hear once. Others, especially the quizzes and memory narratives at the end of each chapter, you'll want to repeat until you master them. I'll begin with some background about how generic and brand drugs differ, how we name them, and how to use that to your advantage.

The middle of this introductory chapter includes some references to organic chemistry and may go over your head on first reading. Some people learn science first, and then pick up a passion for reading and writing. Others like reading and writing first and that inspires their passion for science.

At a weeklong event at my community college every March, we have innovators come in to speak. At a meet and greet after his presentation, I had a chance to talk to Homer Hickam, Jr. who wrote the book *Rocket Boys*, later made into the movie *October Sky*. I asked him how he became such a good writer and speaker. He earned a college degree from Virginia Tech in engineering. He said, "I was a writer first."

In the section, if you find yourself frustrated to the point of wanting to turn the book off, skip ahead to the more accessible next chapter where we visit a pharmacy, then, consider coming back. Let me tell you why I've put in so much detail.

When my mother learned her first programming language, P-L-1, it was an era when job opportunities included only teacher or secretary. She was pregnant with my little brother. My dad was her tutor. You would think that with those circumstances she would want to make her study as efficient as possible.

She did the exact opposite. Each time she learned a concept, she asked, "Why?" Why write the code *that* way? Her attention to those foundational details helped her excel in the class, crush the interview, and get the job.

Pharmacology is a language. This section outlines the etymology, or make-up of drug names to provide you that strong foundation. It answers the question, "why write a drug name *that* way?"

GENERIC NAMES VERSUS BRAND NAMES

From experience, instructors immediately know which names represent *generic* and *brand* drugs. In an academic text, generic names go first in lowercase letters followed by a capitalized brand name in parenthesis. For example, a lowercase "A" starts generic **amphotericin B** and an uppercase "F" starts **the brand name, Fungizone**. In conversation or other writing, students must have the brand or generic name distinction memorized.

Since you don't have the visual clue of the capital letter pointing to a brand name, you will, usually, hear the words "brand," "brand name" or "brand names" immediately preceding a brand name drug.

Some licensing exams only use generic names. I understand; only one generic name exists for each medication. There is one "**acetaminophen**," (with a lower case "A") but we associate the brand **Tylenol** (with an upper case "T") with many products. This *does not* mean a student should *not learn* brand names. A dangerous confusion between generic and brand names is that of **Pepto-Bismol** or **Pepto** for short. Let me tell you a story.

On a web page, I saw a parent post that she gave her 8-year-old child a teaspoonful of **Pepto**. Was that okay? The *Children's* **Pepto** only comes as a tablet. Something was wrong. **Pepto-***Bismol* is the brand name for the generic liquid product **bismuth subsalicylate**. That salicylate is similar to **aspirin**, which is **acetylsalicylic acid** and can cause a terrible condition called **Reye's syndrome** in children. What she meant to give her child was the brand **Children's Pepto**, which contains **calcium carbonate**, the active ingredient in the brand name **Tums**.

I'm a parent. I understand what happens in the middle of the night. Your child is suffering and the medication instructions are in tiny 4-point font. It's tough to make the right call.

Most *brand* names will have two to three syllables. That's the case with the brand **Pepto**. Most *generic* names have four or more syllables. For example, **calcium carbonate** and **bismuth subsalicylate** have six syllables or more. Why does this matter?

You want to develop simple *heuristics*, or rules of thumb, to make it easier to learn the medicines. For example, generic names have stems like the "-cillin" in penicillin class antibiotics; brand names do not. Getting meaning from a generic or from a brand name requires a different rule of thumb. If you don't know if the drug name you are looking at is generic or brand, you are at a terrible disadvantage.

THREE TYPES OF DRUG NAMES

The chemical name, the generic name, and the brand or trade name

Drug name type 1) The chemical name: First, there is the most complex name, the International Union of Pure and Applied Chemistry (I-U-P-A-C) standard name, which makes perfect sense to a chemist who might want to draw the molecule. For example, the generic name **ibuprofen** has the chemical name:

(RS)-2-(4-(2-methylpropyl)phenyl) propanoic acid

That's simply too unwieldy to use in conversation. Chemists may simplify the chemical name. In this example, the chemical name can become the common name, Iso-butyl-propanoic-phenolic acid. Take I from Iso, Bu from butyl, Pro from propanoic, and Phen from phenolic to make I bu pro p-h-e-n. This we can shorten by one letter to ibupro f-e-n, the generic name.

Drug name type 2) The generic name. The transformation to **ibuprofen**'s four syllables improves on the chemical name. Patients, however, prefer two to three syllable names. Just as kids cut down computer to "puter" or banana to "nana," patients prefer short names for easy pronunciation.

If you don't know where to put an emphasis on a generic four syllable drug name, it's usually the penultimate, or second to last. Pronounce the name as **i bu PRO fen**.

You want to make sure to put the right *emphasis* on the right *syllable* or you'll lose credibility. If your patient pushes the call button when you walk *into* the room, you might be mispronouncing some meds.

Drug name type 3) The brand or trade or proprietary name: Two of the brand names for **ibuprofen** are **Advil** and **Motrin**. Both have two easy-to-pronounce syllables, but neither resembles **ibuprofen**. The brand names include phonetic *plosives* to make powerful memorable stops in the word. A strong word sounds like strong medicine and this likely helps with prescription sales.

Say each of the following letters and see if you can feel it in your tongue or nose.

Tongue blade occlusions for the letters *t or d*

Tongue body occlusions for the letters *k or g*

Lip occlusions for the letters *p or b*

Nasal stops for the letters *m or n*

For example, the brand name Motrin has an "m," a nasal stop; a "t," a tongue blade occlusion; and an "n," another nasal stop. This forces the person saying the word to stop their breath three times, slowing the pronunciation and keeping it on the tongue and or nose longer, making it sound strong.

BRAND NAME RULE OF THUMB: CAN INDICATE FUNCTION

The **brand name's** two to three syllables will sometimes hint at the drug's *function*. For example., **Lopressor** lowers blood pressure.

The **brand name** resembles a two- to three-syllable nickname that *hints* at the drug's function. By law, drug names may not make therapeutic claims. Brand names are much like nicknames such as Betsy or Jack. A nonnative English speaker would have no idea that Betsy comes from Elizabeth or Jack comes from Jonathan. Betsy takes the "b-e-t" from Elizabeth and Jack takes the "J" from Jonathan.

You might see brand names do the same. We use "val" in the brand name **Valtrex** (an antiviral), which comes from a part of the generic **val**a**c**y**c**lovir. The "p," "a," and "x" in the brand name **Pax**il (an antidepressant), comes from the generic name **p**a**rox**etine.

GENERIC NAME RULE OF THUMB: CAN INDICATE CLASS

The brand name **Lopressor**'s four-syllable generic name, **metoprolol** has an o-l-o-l suffix, which is a **stem** from a credible source.

I know of two stem lists. One is from the **United States Adopted Names Council** on the American Medical Association's website and another is the **World Health Organization's** (WHO) Stem Book. They created stems so when people use generic names, they'll know the drugs with similar stems are similar in their actions.

Students, recognizing similar endings or beginnings of drug names, will sometimes make up their own shortcuts. We'll discuss which stems are credible and verifiable and which aren't. Knowing both will make you much better at spotting them. Most YouTube videos and Quizlet notecards have errors that let you know the author's research is faulty or nonexistent. I'm not picking on students; many licensed health professionals give advice and charge money online to view their videos, and some of that advice is dead wrong. However, when used correctly, these prefixes, suffixes, and infixes can be invaluable.

Some of the references, online videos, and online pre-made note cards provide lists deriving stems mistakenly. For example, these nine endings -azole, a-z-o-l-e, -en, e-n, -ide, i-d-e, -in, i-n, -ine, i-n-e, -one, o-n-e, -pam, p-a-m, -sone, s-o-n-e, and –zine, z-i-n-e, are *not* stems. Using an unestablished group of letters instead of a credible stem might lead to a drug classification error.

PREFIXES, SUFFIXES, AND INFIXES

I'll use the terms prefix, suffix, and infix. Generic drug names are invented words and do not always conform to the rules of English. The United States Adopted Names Council properly calls each prefix, suffix, and infix that has meaning a stem. For example, "cef- , c-e-f-hyphen," represents cephalosporins and "-cillin, hyphen-c-i-l-l-i-n," represents penicillins and "–pred-, hyphen p-r-e-d hyphen" from methyl*pred*nisolone represents prednisone and methylprednisolone derivatives. I'll stop using the word hyphen just to keep from being repetitious, but know they are there when trying to indicate if a stem is a prefix, suffix, or infix.

In the case of penicillin, -cillin is the stem and peni- is a prefix that differentiates peni*cillin* from other penicillin antibiotics such as amoxi*cillin* or ampi*cillin*.

An infix added to a stem (not an infix with drug classification meaning) is inside the word and makes the classification more specific. This descriptive infix is like "red" as an adjective describing a car's color.

The proper stem for a quinolone antibiotic is "oxacin, o-x-a-c-i-n." But cipro*fl*oxacin has the infix –F-L- to classify it further as a *fluoro*quinolone, one that contains a fluorine atom.

Stems work as heuristics. Stems allow us to speed up our recognition of medication classes. If there is a list of ten medications that end in "olol, o-l-o-l", we can more easily know these meds are beta-blockers.

Let me give you four examples of the stems and the mountain of information they provide. I'll go over four suffixes as an example: –adol, -afil, -amivir, and -azepam.

1) The stem "adol, a-d-o-l" indicates that a drug like **tramadol (brand name Ultram)**, is an analgesic (a medication for pain), comprised of an opiate (which means from the opium poppy. **Tramadol** has properties as both an agonist

(a chemical that stimulates receptors) and an antagonist (a chemical that blocks receptors). Usually chemicals are agonists or antagonists, not both.

2) The stem "afil, a-f-i-l" in **sildenafil (brand, Viagra)** and **tadalafil (brand, Cialis)** represents the phosphodiesterase type-5 (P-D-E-5) inhibitors. These medications block (inhibit) an enzyme (phosphodiesterase). This inhibition stops the breakdown of a chemical in the corpus cavernosum to help patients who have erectile dysfunction, E-D.

3) The stem "amivir, a-m-i-v-i-r" in **oseltamivir (brand, Tamiflu)** and **zanamivir (brand, Relenza)** is a subclass of the stem –vir (hyphen-v-i-r). Amivir represents the neuraminidase antagonist group. Neuraminidase is an enzyme critical for influenza virus replication. If the patient takes the medication within the first 48 hours after symptom onset, it blocks the enzymes necessary for influenza viruses to reproduce.

4) The stem "azepam, a-z-e-p-a-m" represents antianxiety agents in the benzodiazepine class that are similar to **diazepam (brand, Valium)**. Besides anxiety, patients use these medications as a sedative-hypnotic.

I hope you see how much information these stems provide.

PRONUNCIATION AND ORGANIC CHEMISTRY WORD PARTS

Here's the organic chemistry section I mentioned.

Pronouncing generic names is often like pronouncing foreign language last names like mine. My last name, "Guerra," has a part that's unpronounceable in regular English because there is no double or rolled "r" sound in English. Drug names might use the same letters you know in

INTRODUCTION

the Roman alphabet. However, we pronounce them differently because the sounds that form them come from organic chemists and biochemists.

Note: In the following, I have stressed the part of the generic name to which the organic molecule matches. These are *not* classification stems like –cillin or –azepam. These parts reference certain chemicals or groups that chemists might pronounce differently than you. For example, the word "fur" means an animal covering, but in organic chemistry, we pronounce "f-u-r" as FYOOR.

The following words suggest the number of carbon atoms in an attached molecule made up of only carbon and hydrogen atoms:

Methyl – where m-e-t-h-y-l begins *Methyl*phenidate, for Attention Deficit Hyperactivity Disorder   METH-ill

Ethyl – where the final y-l completes Fentan*yl*, an opioid analgesic   ETH-ill

Propyl – where p-r-o appears in the middle of Meto*pro*lol, a beta blocker   PROP-ill

And butyl – where b-u appears in the middle of Al*bu*terol, a bronchodilating beta 2 receptor agonist   BYOOT-ill

Levo and dextro mean left and right respectively:

Here, Levo – l-e-v-o begins *Levo*thyroxine, a thyroid hormone supplement   LEE-vo

And d-e-x from Dextro – begins *Dex*methylphenidate, for Attention Deficit Hyperactivity Disorder   DEX-trow

These next words mean there is a specific chemical element in each molecule:

The t-h-i from Thio indicates a sulfur atom is in Hydrochloro*thi*azide, a diuretic   THIGH-oh

Chloro indicates a chlorine atom is in Hydro*chloro*thiazide, a diuretic   KLOR-oh

Memorizing Pharmacology

Hydro indicates a hydrogen atom is in *Hydro*codone, an opioid analgesic *HIGH-droe*

And the f-l from Fluoro indicates there is a fluorine atom in the molecule and appears in the middle of Cipro*fl*oxacin, a fluoroquinolone antibiotic *FLOR-oh*

In this next list, the words are branches that attach to the central molecule:

Acetyl – where the letters a-c-e-t appear in Levetir*acet*am, an antiepileptic *Uh-SEAT-ill*

Alcohol – employing the final o-l in Tramad*ol*, a mixed opioid analgesic *AL-kuh-haul*

Amide – a-m-i-d-e, which ends Loper*amide*, an antidiarrheal *UH-myde*

Amine – a-m-i-n-e, which ends Diphenhydr*amine*, a first generation antihistamine *UH-mean*

Disulfide – where the first d-i-s-u-l-f begins *Disulf*iram, an alcoholism medication *DIE-sulf-eyed*

Furan – where the first letters f-u-r begin *Fur*osemide, a loop diuretic *FYOOR-an*

Guanidine – using its final i-d-i-n-e to end Cimet*idine*, an H-2 receptor antagonist for excess acid *GWAN-eh-dean*

Hydroxide – employed in its entirety in Magnesium *Hydroxide* HI-*drox-eyed*, an antacid

Imidazole – using the final a-z-o-l-e to end Omepr*azole*, a proton pump inhibitor *im-id-AZ-ole*

Ketone – using the final o-n-e to end Spironolact*one*, a potassium sparing diuretic *KEY-tone*

Phenol – from which the first letters p-h-e-n complete the word Acetamino*phen*, a non-narcotic analgesic *FEN-ole*

And Sulfa – (s-u-l-f-a) which begins the word *Sulfa*methoxazole, a sulfa antibiotic *SULL-fuh*

It's important to know those names come from organic chemistry parts, but what you want to *memorize* are the ways chemists use to classify drugs. Here are four methods:

1. By what the drug does for the patient, also called the therapeutic class, for example, anti-depressant
2. By their chemical structure: Tricyclic antidepressants (T-C-As) have three rings in the chemical compound
3. By the receptor they affect: Beta-blockers block beta receptors
4. By the neurotransmitter they affect: Selective serotonin reuptake inhibitor (S-S-R-Is) affect serotonin

You'll become familiar with these classifications as we progress through the book, so no need to worry if that section made you feel a little uneasy.

That's another reason to discuss medications with others. You'll pick up the pronunciation as you make mistakes or listen to others have slipups. Mistakes are a part of learning. You can't exactly pick up pronunciation from just looking at brand and generic names in a book. That's why I went to so much trouble to create an audiobook. If you pay attention to drug names, you *will* find clues for building strong mnemonics.

HOMOPHONES

In grade school, you may have learned the word "homophone." Homophones are words that sound the same, but that we spell differently. Some examples include the words "their" (t-h-e-i-r) and "there," (t-h-e-r-e) "two" (t-w-o) and "too," (t-o-o) and "hear" (h-e-a-r) and "here" (h-e-r-e). To remember which meaning is associated with the word, your teacher may have told you to look inside the word for a clue.

In "their" (t-h-e-i-r) you find "heir," (h-e-i-r) – a person who inherits. Then you associate "their" (t-h-e-i-r) with a group of people versus "there" (t-h-e-r-e) which means "in that place."

In the word "two," (t-w-o) you can turn the "w" sideways to make a 3, spelling "t-3-oh" to remember that two has to do with a number. Also, "too" (t-o-o) has two o's. Remember it has too many o's.

In the word "hear" (h-e-a-r) you find the word "ear," which reminds you this word means to listen, versus "here" (h-e-r-e) which means "in this place."

Those clues are mnemonic devices, something to help your memory. Mnemonics also work in drug memorization.

Just a random note: The word mnemonic comes from the name Mnemosyne, the goddess of memory in Greek mythology.

MNEMONICS IN PHARMACOLOGY

The brand name **Prilosec**, a proton pump inhibitor for reducing stomach acid, contains "P-r" which can be short for "proton" (H^+), the ion associated with something acidic. **Prilosec** contains "l-o" which is short for "low," the opposite of high. **Prilosec** contains "sec" which can be short for "secretion."

Prilosec's mechanism of action (M-O-A) is to inhibit proton pumps and reduce the acid in a person's stomach. By looking at the name of the drug, we can see that "proton" "low" "secretion" means a decrease in protons, helping us remember the meaning of the brand name.

What if you only memorized **omeprazole**, the generic name? If you forgot the stem, -prazole, or the drug's purpose, you're in trouble. Brand names serve as a back-up plan.

When developing my mnemonics, I didn't call anyone at any brand name drug companies. I looked at each drug name and used my experience as a teacher of pathophysiology, pharmacology, and organic and biochemistry, to make meaning.

I also asked my students, "what do you see here? How would *you* remember this?"

I looked for what would help students remember the drug's class and function, using the methods they used.

The FDA doesn't allow a drug company to name a drug after its primary purpose. But, many drug names hint at that use.

3 BY 5 NOTECARDS

Many students prefer to use notecards rather than just relying on a book. I think you can get everything you need from this book alone, but I know that 3 by 5 notecards you *make* are much better than any you can *buy*. A book from Harvard University Press titled *Make It Stick: The Science of Successful Learning*, by Peter Brown, goes into why generative (making things) learning is so important.

Notecards are portable. You can sort them and challenge others to sort them like UNO or playing cards. To help you along, I have created a specific order that makes sense out of these 200 cards so you know *how* to sort them. The "how to sort them" aspect is the next level of learning beyond memorizing the purpose of drugs or their generic and brand names.

In this book, related drugs sit next to one another within a group or physiologic system. You'll quickly see connections between them. You will learn new *types* of available connections. We will take small steps, but I know you will be impressed when you can name every single drug's brand name, generic name, and class from memory.

Let me describe a sample notecard:

Printed on the front of the 3 by 5 card is famo<u>tidine</u>, where you've underlined the stem tidine, t-i-d-i-n-e.

On the back of the 3 by 5 card, you've printed H_2 blockers have the stem "-tidine." Looks like "to dine," which we can associate with GERD (capital G-E-R-D).

The brand name, **Pepcid** has "pep," p-e-p, from peptic, which means digestion. **Pepcid** has the "cid" from acid.

COMPREHENSIVE DRUG LIST DISCUSSION

I didn't make a list that covers several pages to intimidate you. It's just that if you spread out 200 3 by 5 notecards, 10 down and 20 across, it makes an area 4 feet by 8 feet. This list helps you see the whole forest – connections between drugs and stems unapparent in a stack of cards.

To set up the framework, memorize the seven pathophysiologic classes in this book in order as G-M-RINCE as in, Grand Mothers RINCE kids' hair. Except, it's the French r-i-n-c-e instead of the English r-i-n-s-e. These seven pathophysiologic classes will form the steel reinforcing bars that, when surrounded by concrete, will provide the foundation for your memorizing the (G) gastrointestinal, (M) musculoskeletal, (R) respiratory, (I) immune, (N) neuro, (C) cardio, and (E) endocrine systems' medications.

This G-M-RINCE order places drug classes from easiest to hardest to learn. To make it easier to memorize the whole list of 200, I had to create some simple rules – a computer programmer might call these algorithms. The two major rules are:

1. Each of the seven sections has over-the-counter (OTC) medications presented first and then prescription-only (RX) medications after. I present all OTC gastrointestinal products before all R-X products, so the drugs that students can physically interact with at the pharmacy come first.
2. I alphabetized drugs in the same class unless there is a pharmacologic reason to consider them out of alphabetical order.

For example, **diphenhydramine**, a 1st-generation antihistamine that starts with "d," would go before **cetirizine**, a 2nd-generation antihistamine that starts with "c." The generational move from 1st to 2nd overrides the alphabetical order. However, **cetirizine** and **loratadine** are both 2nd generation antihistamines so those *are* in alphabetical order. Therefore, the order becomes **diphenhydramine, cetirizine, loratadine** – one first-generation and two second-generation antihistamines.

This is how our brains work – ever consolidating and organizing until meaning emerges from a compact list. The eventual goal is to consolidate all seven chapters in the comprehensive drug list.

You shouldn't need 200 notecards to memorize 200 drugs; you should only need 7 – one for each physiologic group.

OTC SCAVENGER HUNT

I want to walk through the over-the-counter aisles at the corner pharmacy. We're fortunate in the United States to have direct access to many over-the-counter medications in multiple places: supermarkets, drugstores, convenience stores and gas stations.

At its heart, pharmacology is a lab science. We won't mix any chemicals today, but we will experience pharmacology kinesthetically – with our hands. Well, they probably won't let us touch the Sudafed stored behind the pharmacy counter.

As a young pharmacy student, my first retail experience was a weeklong rotation at my childhood grocery store. The pharmacist took me to the over-the-counter medications and said, "I want you to learn these first and then come back to the pharmacy after an hour."

I was exhilarated to What did he want? Should I take notes on each drug? Did he want me to memorize them? Where should I start? I was too embarrassed to ask.

Fortunately, at 9:00 in the morning, the aisle was empty and customers only sporadically passed. At that time, I felt much like a pharmacology student on the first day – alone and confused. I

Behind the pharmacy counter, medications sit on shelves in strict alphabetical order. Over-the-counter medications, however, line shelves in their own respective categories by ailment. Someone who looks for an over-the-counter medication zeroes in on a specific target as our brains do.

Think of the word "blue." What happens immediately? You start thinking of blue objects: skies, waterfalls, oceans and, of course, book covers of excellent pharmacology primers like this one. Now, think of the word "stomachache." What comes to mind as we walk into this drugstore?

Right, Pepto-Bismol, Tums, commercials about purple pills. This is how you want to memorize medications. Place them in mental spaces that mirror physical spaces.

To build strong mental visuals, however, we need to *walk* the aisles and ask ourselves, "What stands out?" Our brains provide excellent recall of spaces and images, but textual names elude us. Can you remember your last medicine's exact spelling or the person's name who helped you?

Instead of starting with the daunting goal of memorizing two-hundred drugs, we're just going to interact with these 38 over-the-counter medications in no more time than it would take to sit down to lunch.

No pressure here, let's just do some afternoon shopping.

Listen to me recite these 10 gastrointestinal medications and try to recognize those you've used before. Do you identify with the generic or brand name or both?

1. Calcium carbonate (brand name Tums)
2. Magnesium hydroxide (brand Milk of Magnesia)
3. Famotidine (brand Pepcid)
4. Ranitidine (brand Zantac)
5. Esomeprazole (brand Nexium)
6. Omeprazole (brand Prilosec)
7. Bismuth subsalicylate (brand Pepto-Bismol)
8. Loperamide (brand Imodium)
9. Docusate sodium (brand Colace)
10. Polyethylene glycol (brand MiraLax)

Could you remember these ten medications if I asked you to repeat them back to me? No? Why not?

Your working memory can only hold 4 to 7 items at a time.

We need to *chunk* the information into five pairs by drug class, a manageable amount for your working memory: 1) antacids, 2) H-2 blockers, 3) proton pump inhibitors, 4) antidiarrheals, and 5) laxatives. Let's look at this first category:

1) Antacids

Our new list of five pairs starts with the generic antacids **calcium carbonate** and **magnesium hydroxide**. But, if we want a clerk or cashier to point us to these medications, it might be easier to ask for the shorter brand names **Tums** or

Memorizing Pharmacology

Milk of Magnesia. In this store, **Tums** and **Milk of Magnesia** sit under a white signpost that reads "digestive."

First, notice **Tums** comes as a chalky, colored tablet. **Milk of Magnesia** comes in a viscous white liquid in a cobalt blue bottle with a wave and a curve at the neck. It may seem silly, but take a minute to turn the **Tums** bottle gently from side to side and hear the tablets roll against each. Now, take a minute to hold the **Milk of Magnesia** bottle. Is it heavier or lighter than you thought? Does the liquid move quickly or does it feel like there's syrup inside? Experience medicines with multiple senses to better store them in long-term memory.

For our patients, we need to know both generic and brand names. How do we learn them? Many students use plain white 3 by 5 notecards, brand on front, generic on back, but color is important. Imagine a cherry chewable **Tums** tablet dissolving or the milk-like **Milk of Magnesia** coating your stomach. Actions produce memorable experiences static letters on the page don't.

To reinforce the close relationship between **calcium carbonate** and **magnesium hydroxide**, note the Periodic Table of Elements. Magnesium's "M-g 12" sits directly above calcium's "C-a 20" with comparable chemical properties.

Both medicines work in minutes to relieve acid reflux. They increase the pH and decrease the acidity causing the burning feeling. Note, these two drugs have additional indications. **Tums** supplements calcium and **Milk of Magnesia** has a laxative effect in high enough doses.

Now, can you remember the antacids calcium carbonate, brand **Tums** and magnesium hydroxide, brand **Milk of Magnesia**? Good, let's move to another pair.

2) Histamine-2 Receptor Antagonists

Pepcid A-C, (A-C means acid controller), has a blue box with white lettering and neighbors a group of **Zantac** boxes, also blue with white lettering and a silver shield. The silver shield with a "Z" stamped on it stands to "block acid."

Technically, manufacturers can't say exactly what a drug is for through its brand name, but you can find textual clues.

Pepcid takes the p-e-p from "peptic," which means digestive and the c-i-d from acid. **Pepcid**. This also works in Spanish, if that's your first language, p-e-p from peptico and c-i-d from acido. Still **Pepcid**.

In brand **Zantac**, if you remove the letter zee, or zed as they say in the UK, from Zantac, you find the beginning of the word antagonist and the beginning of the word acid, "ant-ac". **Zantac**.

Turn the boxes around to see the **"-tidine, t-i-d-i-n-e"** ending in **Pepcid's** generic *famo*tidine and **Zantac's** generic *rani*tidine. These are official stems in the suffix position. It means these drugs both belong to the histamine-2 receptor antagonist class. They block stomach acid production.

Quick note: there *is* a histamine-1. It causes allergic conditions we treat with antihistamines like diphenhydramine (brand, Benadryl) and loratadine (brand, Claritin), but we'll get to those in the respiratory section.

3) Proton Pump Inhibitors

What is a proton pump inhibitor?

Protons increase acidity. If we can stop pumping protons into the stomach, we can reduce that acidity.

When you reach the proton pump inhibitors on the shelf, you will notice brightly colored purple and pink packaging. **Nexium** is in a yellow box with a purple tablet in the center, while **Prilosec** is in a purple box with a purple tablet in the center.

Turn the boxes around side-by-side and look to the similarities.

We see that **Prilosec's** generic name is **ome***prazole* and **Nexium's** generic name is *esomeprazole*. I'll talk a little more about the chemistry behind this later, but to find the "stem" we'll look at the back of one more proton pump inhibitor (not on our immediate list), **Prevacid.**

It comes in a pink box and helps "*prev*ent *acid*, taking the P-r-e-v from prevent and using the whole word "acid" to make **Prevacid**.

Prevacid has the generic name **lanso***prazole*. The proton pump inhibitor's all have the common ending–*prazole*, p-r-a-z-o-l-e and we can verify this with the published stem list.

Some of the following over-the-counter drugs also have stems, but I'll put greater focus on those later in the book.

4) Antidiarrheals

We've paired two antacids, two H_2 blockers, and two proton pump inhibitors. Next we'll pair two antidiarrheals: **bismuth subsalicylate, brand Pepto** and **loperamide**, brand **Imodium**.

For patients experiencing an uncomfortable bout of diarrhea, finding an OTC remedy in the aisles quickly can be easy. **Pepto's** bright pink bottle holds an opaque liquid flavored either wintergreen or cherry.

Look at the word Pepto-Bismol. Think of peptic which, again, means digestive. Also think of bismuth as number 83 on the Periodic Table of Elements.

The generic stem in **bismuth subsalicylate** is –sal-, s-a-l, indicating a salicylate component. When we think of salicylates we usually think of treating analgesia with **aspirin**, chemical name **acetylsalicylic acid**.

However, in this case we memorize –sal, s-a-l to remember that children shouldn't take bismuth subsalicylate, just as they should not take aspirin.

Bismuth darkens the tongue and stool, but this is harmless.

Bright green bottles hold another antidiarrheal medication, the mint flavored liquid **loperamide,** brand **Imodium.** **Loperamide** doesn't have a stem within its name, but you can use some word clues. Think of peristalsis, or the movement in your stomach that moves food through your intestines.

If you can think of **lo-per-amide** as low-peristalsis, you can remember it slows bowel movement.

You can also relate brand **Imodium** to the word immobile, to stop things from moving as well.

5) Laxatives

For patients with the opposite problem – constipation, we have one stool softener and one laxative: **docusate sodium, brand Colace** and **polyethylene glycol, brand MiraLax.**

Docusate comes in a small, often difficult to find bottle. As a stool softener, docusate brings water into the bowels to move stools more easily. Often, when pharmacists instruct patients to take a medication with a full glass of water, it's because a large pill with a large volume will move down the esophagus more easily.

Docusate can dehydrate the patient, so we're providing water for rehydration. Even though it doesn't exactly work in the colon, you can think of **Colace** as speeding the colon's pace; **Colace.**

Polyethylene glycol towers above docusate in its large white bottles with dark purple lettering. The colored screw on top serves as a measuring cup for the powder inside. The stem for **polyethylene glycol** is –peg, p-e-g and you can pull those letters from the generic name. It is an osmotic laxative, safe to use in children. The brand name alludes to miracle plus laxative. MiraLax.

Let's go on to the next drug category and listen for familiar products.

5 musculoskeletal drugs

1. Aspirin regular strength, (brand, Ecotrin)
2. Ibuprofen (brand names, Advil and Motrin)
3. Naproxen (brand, Aleve)
4. Acetaminophen (brand, Tylenol)
5. Acetaminophen with aspirin and caffeine (brand, Excedrin Migraine)

Note: Textbooks generally list these medications under the musculoskeletal heading, but in the pharmacy aisle, you will see them under the "analgesics" sign post.

1) Aspirin, brand Ecotrin

Aspirin is a *non*-steroidal anti-inflammatory drug, and NSAID. A *steroidal* anti-inflammatory drug would include names like prednisone or methylprednisolone.

Aspirin's brand name **Ecotrin** hints at its enteric coating Enteric-coated medications pass through the acidic stomach without dissolving until it reaches the more basic pH environment of the small intestine. This works to prevent stomach irritation and ulcers.

Patients take regular 325 milligram strength Ecotrin (often in an orange box with yellow designs) for muscle pain, fever, and other conditions. The low dose 81 milligram strength, (in a blue box), works to prevent stroke and heart attack. Many companies manufacturer **aspirin.** Double check the dosing and watch for this color contrast.

2) Ibuprofen, brand names Advil and Motrin

Ibuprofen is a nonsteroidal anti-inflammatory drug, NSAID, similar to **aspirin**, but safe for children. If you see a patient scouring for children's ibuprofen, they may not realize it's in a different section than the adult medications. Within the children's section, we often differentiate children's and infant's formulations. Asking for a pharmacist's help is appropriate.

While brands **Advil** and **Motrin** contain the same active ibuprofen ingredient, the boxes' colors contrast significantly. Competitors want their product to stand out, so **Advil's** blue box sits exactly opposite of **Motrin's** orange box on the color wheel. These marketing tactics impact our visual literacy and can influence our buying decisions.

3) Naproxen, brand Aleve

We dose **aspirin** and **ibuprofen** up to four times daily. But **naproxen**, brand **Aleve**, only needs twice daily dosing. Think of **Aleve** as *allev*iating your pain. It's also in a blue box, a color we often associate with health.

4) Acetaminophen, brand Tylenol

Acetaminophen is a *non*-narcotic analgesic to contrast medications like morphine that are *narcotic* analgesics. Just as **aspirin, ibuprofen**, and **naproxen** are *non*-steroidals, **acetaminophen** is a *non*-narcotic. It is like defining nonfiction by saying it's not fiction .

Acetaminophen's bright red box with white lettering is easy to spot on pharmacy shelves. The generic name acetaminophen and brand name Tylenol come from parts of the chemical name, N-acetyl-para-aminophenol.

5) Acetaminophen with aspirin and caffeine, brand **Excedrin Migraine**.

Excedrin Migraine combines lower doses of acetaminophen and aspirin with caffeine.

Acetaminophen acts as a non-narcotic analgesic to alleviate migraine pain, but it doesn't help reduce the inflammatory process. A**spirin** helps as an anti-inflammatory and second analgesic.

Why not just buy a separate bottle of aspirin and acetaminophen and combine it with coffee or a caffeinated energy drink?

There is a school of thought that giving two medications allows us to reduce the required dosage to limit side effects.

It seems a little bit strange to include caffeine, doesn't it? Do you want to be wide-awake during your headache?

The caffeine is not for wakefulness. Rather, a headache is thought to be in part, an issue of brain blood vessels dilating. Caffeine constricts, or narrows blood vessels, to relieve headache pain.

Next, we'll review 9 Respiratory Medications. As I list them, again, think of ones you've come in contact with or seen television commercials for.

1. Diphenhydramine (brand Benadryl)
2. Cetirizine (brand Zyrtec)
3. Loratadine (brand Claritin)
4. Loratadine-D (brand Claritin-D)
5. Pseudoephedrine (brand Sudafed)
6. Phenylephrine (brand Neo-Synephrine)
7. Oxymetazoline (brand Afrin)
8. Triamcinolone (brand Nasacort Allergy 24HR)
9. Guaifenesin with dextromethorphan (brands Robitussin DM or Mucinex DM)

1) Diphenhydramine, brand Benadryl

In a film comedy about a dating coach named Hitch, played by Will Smith, there was a scene where he went to the pharmacy's over-the-counter aisle to treat his food allergies.

He looked into the security mirror, saw his severely swollen face, and then knocked over half the medicines on the shelf before drinking a good helping of pink diphenhydramine, brand Benadryl.

Unfortunately, he likely took the wrong medicine, but it was good comedy. An oral liquid works more slowly than an intramuscular injection. A person having a rapid allergic reaction would likely need an epinephrine injection.

The story serves as a reminder patients should get help even though this department allows for self-service.

Diphenhydramine, brand Benadryl, is a first-generation antihistamine. It acts on histamine-1 receptors to reduce common allergy symptoms like sneezing, itching, and hives.

A first-generation drug is the first in a class to show a desired activity. Subsequent generations form a successive line of derivatives from the original drugs, often with fewer side effects due to molecular structural differences.

It's important to know that the sedation first generation antihistamines cause can limit diphenhydramine's use. Patients may be too tired to work and shouldn't drive while on it. But this side effect can also be a positive when used as a sedative in Tylenol PM, for example.

To remember this effect in brand name **Benadryl**, you can see the word "bed" by removing the 'na, n-a' in the middle of the word.

Benadryl contains "ben, b-e-n" at the beginning and "dry, d-r-y" near the end. This can help you remember its benefit comes from drying allergy secretions.

2) Cetirizine, brand Zyrtec

Cast in a striking lime green box, this *second*-generation antihistamine also acts on histamine-1 receptors to reduce allergy symptoms. Unlike diphenhydramine, it produces significantly less drowsiness. Recognize cetirizine may take several days to provide symptomatic improvement whereas diphenhydramine begins working in a few hours.

We spell **cetirizine** with t-i-r in the middle, but pronounce it "tear." Use this clue to remember **cetirizine** can help tearing from allergy eyes.

3) Loratadine, brand Claritin

Loratadine has a recognized stem in its generic name, "-atadine, a-t-a-d-i-n-e" and its brand name is **Claritin**. The catchy marketing slogan, "Get Claritin Clear," helps students remember this drug clears up allergies. **Claritin's** packaging displays a green field and *clear* blue sky. This subtly reminds patients with **Claritin,** seasonal allergies don't have to stop them from enjoying the outdoors.

One confusion patients may have is when to use **loratadine** versus **pseudoephedrine**. With a runny nose we recommend an antihistamine like **loratadine** to dry up secretions. With stuffy noses patients will want to use a decongestant to help relieve sinus pressure associated with congestion.

4) loratadine with pseudoephedrine, brand Claritin-D

For patients with both symptoms, combination **Claritin-D** may be a good therapy option. **Loratadine** is the second-generation antihistamine in this product, and the "D" in **Claritin-D** represents the decongestant **pseudoephedrine.**

While looking at the pharmacy shelf in the allergy aisle, you may see advertisements for **Claritin-D**, but no physical boxes to purchase. You'll find **Claritin-D** located *behind* the pharmacy counter, but available without a prescription. You might ask, "Why would pharmacists store a product for a stuffy nose behind the counter?"

The Combat Methamphetamine Epidemic Act of 2005 set rules for **pseudoephedrine** sales. Customers can only purchase a limited amount per day and during a thirty-day period. **Pseudoephedrine** is a base ingredient in the illicit drug methamphetamine or "meth."

Customers can purchase the decongestant separately from antihistamine, so let's look at pseudoephedrine alone.

5) Pseudoephedrine, brand Sudafed.

Pseudoephedrine is a sympathomimetic drug. What does that mean? Well, if we break down the word, "sympath" refers to the sympathetic nervous system responsible for our fight or flight response. The "mime" means to mimic or copy.

So as a sympatho-mimetic drug, pseudoephedrine creates a similar response to the sympathetic nervous system. It constricts blood vessels in the nasal mucosa to helps relieve sinus pressure, but it also constricts blood vessels throughout the body.

This is why patients with uncontrolled high blood pressure should not use decongestants like **Sudafed**. Some patients will feel jittery and wide-awake while using the product and shouldn't take it close to bedtime.

One student came up with a **Sudafed** mnemonic by saying, "I'm fed up with nasal congestion."

6) Phenylephrine, brand Neo-Synephrine

Phenylephrine is an intranasal product as brand Neo-Synephrine. It's also an oral product ingredient that relieves sinus congestion denoted by the abbreviation "P-E" at the end of the product name.

Phenylephrine can cause the same side effects as **pseudoephedrine,** but is often less effective at relieving symptoms of congestion and doesn't have purchasing restrictions.

7) Oxymetazoline, brand Afrin.

Oxymetazoline's primary use is a very short-term, but immediate solution for nasal congestion. **Afrin** can *cause* rebound congestion if you use it for more than three days in a row.

8) Triamcinolone, brand Nasacort Allergy 24 hour.

Triamcinolone is a corticosteroid, which means it can reduce inflammation. We use it prophylactically to avoid allergy symptoms that are a chronic problem for patients.

The brand name has some textual clues. "Nasa" for the nose, "'cort" to say it's a corticosteroid, "Allergy" to plainly state what symptoms it treats, and "24 hours" to tell us how long it lasts.

Not all respiratory ailments stem from allergies. Chest congestion and coughs are commonly result from viral infections that can't be treated with antibiotics. In these patients, sometimes our only option is to use an expectorant and or cough suppressant.

9) Guaifenesin with dextromethorphan (brands Robitussin DM or Mucinex DM)

The combination product **Robitussin-DM** contains **guaifenesin** with **dextromethorphan.** Some students get 1) expectorants and 2) antitussives confused so let's look a little further.

1) Expectorants and mucolytics act as synonyms. These drugs break up mucus stuck in the chest. **Guaifenesin** is a commonly used over-the-counter expectorant. **Guaifenesin** is available individually as a liquid as **Robitussin** or in tablet form as **Mucinex**.

2) Antitussives help calm or stop a cough. **Dextromethorphan** is an antitussive available over-the-counter represented by the "D-M" in **Robitussin-DM.**

Combining the medications breaks up the mucous in the chest and stops the cough.

Four Immune Medications
1. Neomycin with Polymyxin B and Bacitracin (brand Neosporin)
2. Butenafine (brand Lotrimin Ultra)
3. Influenza vaccine (brand Fluzone)
4. Docosanol (brand Abreva)

Turn the corner to the next aisle. First aid products and creams line the wall. There aren't many antibiotics available

INTRODUCTION

over-the-counter, but when supplied and administered topically they are relatively safe.

A common topical OTC antibiotic formulation is **neomycin, polymyxin B** and **bacitracin** or brand name **Neosporin**.

Butenafine, brand **Lotrimin Ultra** is a topical antifungal product.

The **influenza vaccine**, brand **Fluzone** prevents infection with influenza virus.

Docosanol, brand **Abreva** treats fever blisters and cold sores.

1) Neomycin with Polymyxin B Bacitracin (brand Neosporin)

Above boxes of superhero and princess Band-Aids is a brightly colored golden box labeled **Neosporin**. If I listed the generic components: **neomycin, polymyxin B** and **bacitracin,** you might not recognize them right away. But by taking pieces of each generic antibiotic name you can form the brand **Neosporin**. First, take 'n-e-o-s' from **neomycin sulfate** the "p-o" from **polymyxin** and the "r-i-n" form **bacitracin**. **Neosporin** ointment covers minor cuts and scrapes to help prevent infection.

2) Butenafine, brand Lotrimin Ultra

This antifungal cream sits among the treatments for athlete's foot and ringworm. The brand name for **butenafine** is **Lotrimin Ultra**. If the product is simply labeled **Lotrimin,** it contains the generic antifungal **clotrimazole**. Both are over-the-counter antifungal creams.

This marketing can confuse patients looking for specific ingredients and why it's so important to read the back of OTC boxes.

3) Influenza vaccine (brand Fluzone)

To prevent influenza infection, the Centers for Disease Control and Prevention (C-D-C) recommend many patients receive an annual **influenza vaccine**. This is a prophylactic measure to prevent viral respiratory infections caused by annual strains of influenza.

While vaccines aren't available on the shelves over-the-counter, pharmacists in many states in the U.S. provide them without a written prescription. While adults do not likely need a prescription for an influenza vaccine, depending on the age of a child they may or may not need a prescription from a physician when receiving the vaccine at the pharmacy.

4) Docosanol (brand Abreva)

Docosanol speeds recovery from bothersome cold sores. Finding this product often presents a challenge. It's sometimes highlighted at the end of the aisle on leftover shelf space. The blue tube is smaller than your pinky and comes in a hard plastic clear package.

I wondered, "who would spend roughly $20 on such a small tube?" But then I thought to myself, "well if I was going to a homecoming dance and it's the only time I ever get to attend, then I would probably pay the money for it."

With this idea I thought of **docosanol** and going to the ball.

The brand name **Abreva** is a little easier to remember. It helps abbreviate or shorten the impact of a cold sore.

Docosanol stops an acute viral infection, the next product prevents it.

Four Neuro medications

1. Benzocaine (brand Anbesol)
2. Lidocaine (brand Solarcaine)
3. Meclizine (brand Dramamine)
4. Acetaminophen PM (brand Tylenol PM)

INTRODUCTION

Topical **benzocaine** and **lidocaine** are local anesthetics available without a prescription. **Meclizine** treats dizziness and prevents motion sickness. **Tylenol P-M** helps insomniacs.

1) Benzocaine (brand, Anbesol)

I want to review the difference between anesthetics and analgesics. Anesthetics like **benzocaine** cause numbness and other drugs in the anesthetic family can cause loss of consciousness. Analgesics reduce or eliminate pain.

The now illegal cocaine, an original anesthetic, provided the -caine stem. Once scientists recognized cocaine's addictive properties, they developed newer non-addictive agents like **benzo*caine*** and **lido*caine***.

One of my students gave me a trick to remember the brand name for **benzocaine, Anbesol, A-n-b-e-s-o-l** If you take the "u" and the "m" out of the word numb, n-u-m-b, you are left with the "n" and "b" in **Anbesol**. At the pharmacy, patients can find this small tube with the toothpaste or near other topical oral products. Because **Anbesol** is a topical anesthetic, it aims to numb sore teeth, mouth sores, or gum pain.

2) Lidocaine, brand Solarcaine

Lidocaine also has the -caine stem placing it as an anesthetic. Brand *Solarcaine* help numb pain from sun or solar burning. This product comes as a lime green gel or aerosol can to avoid rubbing sensitive skin.

3) Meclizine, brand Dramamine

Meclizine is actually an antihistamine, but it also relieves motion sickness. Patients going on ocean cruises often ask for it. **Meclizine**, m-e-c-l-i-z-i-n-e, contains all but the "t" from emetic, e-m-e-t-i-c in it to help you think of nausea.

Also, you can connect the lowercase 'c' to the lowercase 'l' to form a lowercase 'd', to spell out 'dizi.'

4) acetaminophen with diphenhydramine, Tylenol PM.

Pain keeps many patients awake. Acetaminophen can control pain and diphenhydramine produces drowsiness. An effective combination.

Three Cardio medications
1. Omega-3-Fatty Ethyl esters (brand Fish Oil)
2. Niacin (brand Niaspan)
3. Aspirin (Low Dose) (brand Ecotrin Lo-Dose)

Physicians manage most cardiology therapies with prescription products. However, two over-the-counter supplements can help lower cholesterol and one medication acts as an antiplatelet. We will start with **Omega-3 Fatty Ethyl Esters**. The ethyl esters are actually a prescription product, but less expensive **Fish Oil** contains **Omega-3 Fatty Acids** over-the-counter. **Niacin** helps lower cholesterol and a daily 81 milligram **low-dose aspirin** reduces stroke and heart attack incidence.

1) Omega-3-Fatty Ethyl esters (brand Fish Oil)

Omega-3 Fatty Acids are an essential dietary fat. They help form cell membranes lower cholesterol levels ultimately reducing heart disease risk. When looking for **Omega-3 Fatty Acids** in the vitamin aisle, look toward the beginning of the alphabet under **Fish Oil**. **Omega-3 Ethyl Esters** are known as prescription **Lovaza**. Lovaza has undergone rigorous F-D-A testing for purity.

2) Niacin (brand Niaspan)

Niacin is nicotinic acid or vitamin B-3. It is supposed to improve cholesterol, increase HDL, and decrease triglycerides. However, **niacin** has a common irritating side effect – the niacin flush. Patients taking this supplement get flushed and rosy cheeks. They limit this adverse effect by taking a 325mg aspirin 30 minutes before.

Niacin works as a relatively inexpensive option for some patients.

3) Aspirin (Low Dose) (brand Ecotrin Lo-Dose)

Aspirin at a low dose of 81mg has an antiplatelet effect as opposed to 325 milligram's analgesic effect.

Aspirin is especially important in patients with blood vessels narrowed from atherosclerosis, or fatty deposits in the arteries. The formation of a clot in these areas could lead to vessel blockage ultimately causing a stroke or heart attack. Patients at risk should take low dose **aspirin** daily to reduce the risk of morbidity and mortality.

Three Endocrine Medications
1. Regular insulin (Brand Humulin R)
2. NPH insulin (Brand Humulin N)
3. Levonorgestrel (Brand Plan B One-Step)

In the final endocrine section we will discuss two insulins, **Regular** and **NPH**, used for diabetic patients. These are available without a prescription in case patients cannot afford a prescriber visit. Patients also request insulins over-the-counter without prescription for their pets. We will also discuss **levonorgestrel**, an over-the-counter emergency contraceptive that can prevent pregnancy.

To understand where **Regular Insulin** and **NPH Insulin** fit on the timeline of insulins you need to know about the others. Humalog is a rapid-acting insulin administered just before a meal and responds quickly – about 15 minutes and lasts about 3-5 hours. It is prescription-only because of a greater risk for hypoglycemic events. Patients must eat with this treatment.

Humulin R or **Regular insulin** has an onset of about 30 to 45 minutes, and works for up to 12 hours. **NPH Insulin** is intermediate-acting with onset around 3 hours, and works for about 18 hours. Lastly, long-acting insulin like **Lantus** starts working within 3 hours, but lasts up to 24 hours.

Even though Regular insulin and NPH insulin are available without a prescription, they will physically be behind the counter in a pharmacy refrigerator. The insulin comes in prefilled vials inside a sealed white box, so that you know no one is tampered with it.

Patients who purchase insulin this way may also request syringes subcutaneous injection administration. Cash-pay patients do not require prescriptions for syringes either, but they are often held in the pharmacy for safety reasons.

The final medication in the endocrine section is **levonorgestrel**, or **Plan B One-Step**. Plan B is an emergency contraceptive that reduces the chance of pregnancy after unprotected sex. It's brand name comes from the thought that plan A failed.

Even when people practice safe sex, and use appropriate contraception, like a condom, they are not 100% effective.

Why is it called One Step? It used to be a two tablet regimen now simplified to only one.

CHAPTER 1: GASTROINTESTINAL

Gastrointestinal Section I. Peptic ulcer disease

What is peptic ulcer disease and how do we treat it?

Clinicians diagnose peptic ulcer disease when an ulceration of the peptic or digestive tract occurs. Acid is an aggressive factor in the stomach that, if lessened, allows an ulcer to heal.

In this chapter, we'll learn about three classes of acid reducers: antacids, histamine$_2$ receptor antagonists (also known as H$_2$ blockers), and proton pump inhibitors, (P-P-Is).

I've ordered the first three pairs of acid reducers from the fastest to slowest working. Antacids take a few minutes, H$_2$Blockers take about half an hour, and proton pump inhibitors can take as long as a day.

Note: We need antibiotics to wipe out *Helicobacter pylori*, the organism responsible for the ulcers, but we will tackle those medications in chapter 4.

Let's start with two antacids.

Antacids, or, anti-acids, can contain the elements **calcium** and **magnesium** to raise the stomach's pH.

The *pH scale* is like a one-foot long ruler with 14 lines instead of 12.

A "0" sits left for the most acidic compounds and a "14" sits right for the most basic or alkaline ones. 7, is in the middle, and represents a neutral state.

From left to right on our ruler, stomach acid has a pH of around 2; milk is about 6; neutral is 7; and human blood is approximately 7.35.

Let's ask ourselves a question. Why do people give dairy milk, which is slightly *acidic*, to calm an *acidic* stomach?

Answer: Six is more basic than two and this raises the stomach's pH.

These antacids also have *additional* benefits. Besides acting as an antacid, **calcium carbonate** can supplement calcium in a diet and **magnesium hydroxide** works as a laxative.

Unfortunately, these compounds are not without their problems. Both antacids can chelate or bind with antibiotics like **doxycycline** and **ciprofloxacin** and make those antibiotics ineffective.

Let's look at some ways to remember two of the most common antacids on the U.S. market.

**1) Calcium carbonate
(brand names Tums, and Children's Pepto)**
CAL-see-um CAR-bow-nate (TUMS)

Students associate the brand name **Tums** with the word tummy to remember it's an antacid.

Calcium carbonate (brand, Children's Pepto) and **bismuth subsalicylate (brand, Adult Pepto-Bismol)** have different ingredients and shouldn't be interchanged. Remember, this is because the salicylate ingredient in the Adult Pepto can cause Reyes syndrome.

2) Magnesium Hydroxide (brand, Milk of Magnesia)
mag-KNEES-e-um high-DROCKS-ide (MILK mag-KNEE-shuh)

Milk of Magnesia looks like dairy milk, which some people still use to calm acidic stomachs.

Put together the "milky texture" and the diarrhea of lactose intolerance to remember its laxative effect.

CHAPTER 1: GASTROINTESTINAL

What will happen if antacids don't work? Patients can choose between two other acid reducer classes: the histamine 2 receptor antagonists or the proton pump inhibitors.

HISTAMINE 2 RECEPTOR ANTAGONISTS (H 2 R As)

When someone says, "I need an antihistamine," they're looking for relief from allergic symptoms like sneezing, runny nose, and watery eyes. Those allergy antihistamines affect histamine *one* (H_1) receptors. We'll cover those in the respiratory chapter.

The term H *two* blocker (more formally, H_2 receptor *antagonist*) stands for histamine receptor type two (H_2) blocker. Histamine two causes acid *to form*.

Blocking histamine two receptors *reduces* acid production.

There are four H2 blockers on the market; we'll go over two of them. You'll notice their generic names, **ranitidine** and **famotidine** both end in "tidine, t-i-d-i-n-e." Related drugs, as we've mentioned earlier, often have related stems. Because tidine is at the end of the word, we call these stems suffixes. Let's look more closely at these two drugs' names.

1) Famo<u>tidine</u> (brand, Pepcid)
Fa–MOE–ti–dean (PEP-sid)

Famotidine has the "-tidine," t-i-d-i-n-e stem indicating it's an H_2 blocker, but how do we remember what the brand name Pepcid is for?

Pepcid contains "pep" from "peptic" and "pepsin" an enzyme related to digestion plus the "cid" from "acid."

One student mentioned Pepsi, the soda, has a pH of 2.4, which is *very* acidic. She said it always gave her heartburn, so that's how she remembered *Pep*cid.

Also, soda pop's "p-o-p" and **Pep**cid's "p-e-p" differ by only one letter.

Memorizing Pharmacology

2) Ranitidine (brand, Zantac)
ra-NI-ti-dean (ZAN-tack)

One student said **ranitidine's** "-tidine" stem looks like "to dine," the time patients might experience gastroesophageal reflux disease (GERD).

The brand **Zantac** has a "Z" that looks like a "2" with "antac" after it, so it's an H *two* blocker, working as an <u>ant</u>-agonist to <u>ac</u>-id.

Putting the 2 with "ant-ac" helps us remember Zantac is an H 2 blocker.

PROTON PUMP INHIBITORS (PPIs)

Proton pump inhibitors block a pump that introduces protons into the stomach, thus making it less acidic. Prescribers combine the P-P-Is with antibiotics for ulcer triple therapy to kill *H. pylori*.

While **esomeprazole** and **omeprazole** have the same "prazole" ending, notice the only difference between **omeprazole** and **esomeprazole** is an "es," (e-s).

Many drugs have chemical structures that have mirror images called *enantiomers*. Instead of calling them right- and left-handed, we call them "R" and "S" from the Latin words *rectus* (right) and *sinister* (left).

In this case, the "S" form is more active biologically.

However, if we put an "s" in front of omeprazole, a patient would have to pronounce it "some-prazole."

The "e-s" allows for a separation between the "S" sound and the compound name, to force us to say it as a chemist would pronounce it.

Let's look more closely at these two compounds.

1) Esomeprazole (brand, Nexium)
es-oh-MEP-rah-zole (NECKS-see-um)

CHAPTER 1: GASTROINTESTINAL

The "prazole, p-r-a-z-o-l-e" ending helps us remember it's a proton pump inhibitor, and the "e-s" means "S" for sinister or originating from the left. But how does the name Nexium relate to hyperacidic states? The manufacturer released **Nexium** after **Prilosec** as the "ne̱xt" PPI drug.

2) Ome<u>prazole</u> (brand, Prilosec)
oh-MEP-rah-zole (PRY-low-sec)

Just as with esomeprazole, the "prazole" ending in **omeprazole** indicates a proton pump inhibitor.

We remember **Prilosec**'s relation to acid reduction by cutting the brand name into three parts.

The "Pr" (p-r) stands for hydrogen p̱rotons (protons, or hydrogen ions, are what make an acid acidic).

The "lo" (l-o) is for lo̱w, the opposite of high.

The "sec" (s-e-c) stands for proton se̱cretion.

Put those three parts together and you get "protons low secretion."

Another way to remember Prilosec's function, and maybe a little less demanding, is to read "Pril oh sec" as "Pril-ZERO-sec" and remember it provides zero heartburn.

GASTROINTESTINAL SECTION II. DIARRHEA, CONSTIPATION, AND EMESIS

Diarrhea can lead to dehydration and sometimes we need to intervene and use over-the-counter medications like **bismuth subsalicylate** or **loperamide**.

Constipation originates from some medication classes.

Opioids like **morphine** decrease gastrointestinal tract motility.

Calcium channel blockers like **verapamil** block calcium from getting to the bowel's smooth muscle slowing peristalsis.

Prescribers frequently give a stool softener like **docusate sodium** and a stimulant laxative for constipation caused by those medications.

Emesis or vomiting is an effective natural body response to ingested toxins. However, with cancer chemotherapy, we want to prevent chemotherapy-induced nausea and vomiting (C-I-N-V) with a drug like **ondansetron.**

Manufacturers specially formulate these antiemetic meds because nausea patients may vomit oral drugs. For example, **ondansetron** comes as an orally disintegrating tablet (O-D-T) that patients take without water.

Promethazine comes in a rectal suppository form.

As with the acid reducers, we'll look at each of these medications in pairs.

Two Antidiarrheals

1) Bismuth Sub*sal*icylate (brand, Pepto-Bismol)
BIZ-muth sub-sal-IS-uh-late (pep-TOE BIZ-mol)

Bismuth's stem is the "sal" in "sub*sal*icylate" and similar to **aspirin**, chemical name acetyl*sal*icylic acid.

Both are dangerous to young children because of the risk of **Reye's syndrome**, a condition involving brain and liver damage that can occur in children with chicken pox or influenza who take **salicylates**.

The "b" in **bismuth subsalicylate** reminds students of the black tongue and black stool that some patients experience as side effects. (Note: This discoloration is harmless.)

The brand name **Pepto** looks like peptic, which has to do with digestion.

2) Loperamide (brand, Imodium)
Low-PER-uh-mide (eh-MOE-dee-um)

The "lo" (l-o) for s<u>lo</u>w and "per" (p-e-r) for <u>per</u>istalsis is how one student remembered *lo*p*er*amide's function.

Imodium is like the word "immobile" in that it slows down the bowel or immobilizes it.

TWO MEDICATIONS FOR CONSTIPATION

1) Docusate Sodium (brand, Colace)
DOCK-you-sate SEWED-e-um (CO-lace)

Docusate sodium softens the stool and patients use it with opioids like **morphine.**

Docusate and "penetrate" rhyme to remind us that **docusate sodium** works by helping water *penetrate* into the bowel.

The brand name **Colace** alludes to improving the <u>col</u>on's p<u>ace</u>. Here we can use the c-o-l from "colon's" and a-c-e from "pace" to construct the brand name, Colace.

2. Polyethylene Glycol (brand, MiraLax), an osmotic laxative
pa-Lee-ETH-ill-een GLY-call (MIR-uh-lacks)

I remember **polyethylene glycol's** function because I have triplets, "**<u>poly,</u>**" call "**<u>c-o-l</u>**" me when they finish going to the bathroom. I often hear, "Daaaaaad! Can you wipe me? *just* as I'm about to sit down to dinner"

Some remember **MiraLax** is the <u>Mira</u>cle <u>Lax</u>ative because it's a miracle how good you feel after taking it.

MiraLax has a "by prescription only cousin."

Go-Lytely is a 4-liter plastic bottle of **polyethylene glycol** used for colonoscopy examination preparation. There is nothing "lightly" about its laxative effect, so you will want to make sure there is a bathroom nearby.

Two Antiemetics

1) Ondan<u>setron</u> (brand names, Zofran, Zofran ODT)
on-DAN-se-tron (ZO-fran)

The "setron" s-e-t-r-o-n suffix will help you remember **ondansetron** is a serotonin 5-HT$_3$ receptor antagonist for preventing emesis.

If you are good at word scrambles, **ondansetron** has every letter but the "i" in serotonin, a neurotransmitter, the majority of which is located in the GI tract.

The "O-D-T" in Zofran O-D-T stands for orally disintegrating tablet. It is a useful dosage form because it dissolves on the top of the tongue and requires no additional liquid.

2) Promethazine (brand, Phenergan)
pro-METH-uh-zeen (FEN-er-gan)

Promethazine is technically an antihistamine manufacturers sometimes combine in liquid form with codeine.

Promethazine also reduces nausea. In addition to oral, IM, and IV forms, **promethazine** comes in a rectal suppository if a patient can't take anything by mouth (po).

Gastrointestinal Section III.
Gastrointestinal Autoimmune Disorders

Autoimmune diseases like ulcerative colitis (U-C) occur when the body's immune system inappropriately attacks an area (or areas) of the body. Symptoms of U-C include ulcerations and inflammation (-itis) in the colon. **Infliximab (brand, Remicade)** blocks the tumor necrosis factor, alpha (T-N-F-alpha), to treat this disease.

Inf<u>liximab</u> (brand, Remicade) *in-FLIX-eh-mab (REM-eh-cade)* is a biologic agent, a genetically engineered protein.

CHAPTER 1: GASTROINTESTINAL

Break up **infliximab's** generic name as (i-n-f) + (l-i) + (x-i) + mab (m-a-b).

The "i-n-f" is a prefix that separates it from other similar drugs.

The "l-i" stands for i̱mmunomodul̲ator (the target).

The "x-i" stands for chimeric, the source. For example, combining genetic material from a mouse, with genetic material from a human.

The "x" might also refer to the Greek letter "chi," which looks like an x.

The "mab" (m-a-b) stands for m̲onoclonal antib̲ody.

Conditions like ulcerative colitis can go into remission so you'll want to think of **R**em**i**c**ade** as a "remi̱ssion ai̱de̱," where the letters r-e-m-i from "remission" and a-d-e from "aide" are prominent in the brand name; **Remicade**.

Chapter 1 Quiz: Gastrointestinal, Drug stem and Drug classification practice

I'm going to read a generic name for you and I want you to think about the stem, if any, and the class of medication.

Calcium carbonate no stem, antacid

Magnesium hydroxide no stem, antacid

Famotidine tidine t-i-d-i-n-e, H two blocker

Ranitidine tidine t-i-d-i-n-e, H two blocker

Esomeprazole prazole p-r-a-z-o-l-e, proton pump inhibitor

Omeprazole prazole p-r-a-z-o-l-e, proton pump inhibitor

Bismuth subsalicylate sal, s-a-l, for salicylic acid derivative which helps you memorize the adverse effect of Reye's syndrome, but its classification is as an antidiarrheal

9

Memorizing Pharmacology

Loperamide no stem, antidiarrheal

Docusate sodium no stem, stool softener

Polyethylene glycol peg, p-e-g, PEGylated compound, laxative

Ondansetron setron s-e-t-r-o-n, anti-emetic

Promethazine no stem, anti-emetic

Infliximab liximab, l-i-x-i-m-a-b, monoclonal antibody for ulcerative colitis

GI: MEMORIZING THE CHAPTER

Now that you have had a chance to get to know each of the drugs in the list individually, you are ready to memorize the first thirteen medications in order. We will connect these 13 gastrointestinal medications to the 25 in the musculoskeletal chapter, the 21 in the respiratory chapter and so on until we've connect *all* 200 drugs. Your brain will be the train's engine that tows these 7 rail cars behind it.

I have included the connections *I made* to help you in memorizing them, but if you have created better ones, use those.

The actor Daniel Radcliffe, of Harry Potter fame, rapped the Alphabet Aerobics on *The Tonight Show Starring Jimmy Fallon*. The Alphabet Aerobics is an incredibly difficult series of words in a specific order.

I hope to one day see a YouTube video with one of my students reciting these 200 medications from memory while another student holds the 200 large drug cards behind them.

I digress. Let's put this all together.

"G" Gastrointestinal

Broadly speaking, there are 13 drugs in the gastrointestinal section, six pairs and a single agent.

Picture in your mind where in the human body these work. Start with the six acid reducers in the stomach, move down to the where the two laxatives and antidiarrheals work in the intestinal tract, and go back up to the antiemetics to prevent vomiting from the mouth, and back down to the colon to the ulcerative colitis medication.

Why this up and down and up and down? It's easiest to start at the stomach and work down to the intestines with O-T-C drugs, then work top to bottom from the mouth to the intestines with the prescription only drugs.

Remember the antacids works the quickest followed by the H2 blockers and the proton pump inhibitors and we'll memorize them in that order.

The antacids **calcium carbonate** and **magnesium hydroxide** are both in the same column on the Periodic Table of Elements in alphabetical order. Two H$_2$ Blockers: **famo<u>tidine</u>** and **rani<u>tidine</u>** follow. Two PPIs **esome<u>prazole</u>** and **ome<u>prazole</u>** follow them. Often diarrhea follows an upset stomach, so we treat with **bismuth sub<u>sali</u>cylate** and **loperamide**. Use "L" from loperamide to get to "L" for two laxatives: **docusate sodium** and **polyethylene glycol**. Use the "p-o" from polyethylene reversed as "o-p" for **ondan<u>set</u>ron** (o) and **promethazine** (p). You can rectally administer promethazine next to the colon to get to the ulcerative colitis medication – **in<u>flix</u>imab**.

Memorizing Pharmacology

Now, can you recite these in order?:

Calcium carbonate Magnesium hydroxide Famo<u>tidine</u> Rani<u>tidine</u> Esome<u>prazole</u> Omeprazole	Bismuth sub<u>sa</u>licylate Loperamide Docusate sodium Polyethylene glycol Ondan<u>setron</u> Promethazine	Infliximab

Two more questions: Do you recognize the stems –tidine, -prazole, -sal-, -setron, -liximab *and* the therapeutic classes they represent?

Listen for them as you hear the list one more time:

Calcium carbonate, magnesium hydroxide, both famotidine and ranitidine with the stem -tidine, both esomeprazole and omeprazole with the stem -prazole, bismuth subsalicylate with the infix –sal-, loperamide, docusate sodium, polyethylene glycol, and the antiemetic names ondansetron with the stem –setron, promethazine, and finally infliximab with its stem, -liximab.

Once you're able to recite the list of thirteen drugs, and you've been able to answer the two questions about stems and therapeutic classes for this chapter, you are ready to move on.

CHAPTER 2: MUSCULOSKELETAL

Musculoskeletal Section I. NSAIDs and Pain

After you've memorized the GI chapter's 13 drugs, the first drug you will connect in the musculoskeletal system as number 14 is aspirin and its related compounds ibuprofen and naproxen that will file into the number 15 and 16 spots.

Non-steroidal anti-inflammatory drugs (NSAIDs) contrast with the steroidal medications used to treat inflammation. The NSAIDs **aspirin** and **ibuprofen,** taken up to four times daily, and **naproxen,** taken twice daily, are available over-the-counter and are for sporadic or mild pain.

Analgesics relieve pain. *Antipyretics* reduce fever. NSAIDs are both analgesics *and* antipyretics.

A unique therapeutic effect of some NSAIDs like **ibuprofen** is that they can close an arterial shunt (*patent ductus arteriosus*) in a preemie. Our newborn daughter had this condition and it caused her to struggle to get enough oxygen.

I watched a YouTube video on the surgery to close the shunt and it took only five minutes, but surgery on any NICU neonate runs a great risk. We were really thankful a simple NSAID closed the shunt and helped her breathe normally again.

"Should I take acetaminophen or aspirin?" is a regular pharmacy question.

Memorizing Pharmacology

If the patient has inflammation, prescribers use NSAIDs, as the non-narcotic analgesic like **acetaminophen** will not help.

However, if the patient has pain or fever, then either is appropriate.

Pregnant patients do not use NSAIDs, rather, they use **acetaminophen**.

For *headaches*, the **brand Excedrin Migraine** employs three drugs: **aspirin** for pain and inflammation, **acetaminophen** for pain, and **caffeine** as a potent vasoconstrictor. Caffeine narrows swollen brain vessels.

Let's look at two prescription NSAIDS.

1) Meloxicam, a prescription only NSAID, has the longest half-life and patients take it once daily. It is like aspirin, ibuprofen, and naproxen in that it blocks both cyclooxygenase-1 and -2, abbreviated COX-1 and COX-2. COX-1 blockade reduces inflammation in the body and, unfortunately, the body's natural protection of the stomach lining.

2) A selective COX-2 inhibitor, like **celecoxib** with the –coxib stem, does not block the protective effect of COX-1 against stomach ulcers and this is why it's supposed to be a better choice.

Let's look at the drugs' names.

THREE OTC NSAID ANALGESICS

1) Aspirin (brand, Ecotrin)
AS-per-in (ECK-oh-trin)

The acronym for **aspirin**, "A-S-A," comes from the chemical name: Acetylsalicylic Acid, taking the "A" from acetyl, S from salicylic, and A from acid.

The brand **Ecotrin** is <u>e</u>nteric-<u>co</u>ated aspi<u>rin</u>, taking the "e" from "enteric," the c-o and t from "coated," and the r-i-n from "aspirin" to make **Ecotrin**.

2) Ibu<u>profen</u> (brand names, Advil, and Motrin)
eye-byou-PRO-fin (ADD-vil, MO-trin)

Many students try to say that **ibuprofen** and **naproxen** both end with "e-n" and that's a good way to remember their relationship as two NSAIDs.

However, many drugs end with "e-n," so that won't help in a large multiple-choice exam.

A better mnemonic is to notice that "profen, p-r-o-f-e-n" from **ibuprofen** is a recognized stem and differs from "proxen, p-r-o-x-e-n" in **naproxen** by only one letter.

3) Naproxen (brand, Aleve)
nap-ROCKS-in (uh-LEAVE)

While **naproxen** has no formal stem, a student came up with a mnemonic for the brand name: **Aleve** will <u>allev</u>iate pain from strains and sprains.

A NON-NARCOTIC OTC ANALGESIC

Acetaminophen (brand, Tylenol)
Uh-seat-uh-MIN-no-fin (TIE-len-all)

The brand **Tylenol**, generic **acetaminophen**, and acronym **A-PAP** all come from the chemical name:

N-ace<u>tyl</u>-para-amino-ph<u>enol</u>

From which we take the t-y-l from "acetyl" and the e-n-o-l ending from "phenol" for the brand name **Tyl-enol**

We take most of the words "acetyl," "amino" and "phenol" for the generic name, **Acetaminophen**.

And we take the four first letters for the acronym, A-P-A-P, or A-PAP.

OTC Migraine - An NSAID with a Non-narcotic analgesic and vasoconstrictor

Aspirin with Acetaminophen and Caffeine (brand, Excedrin Migraine)
AS-per-in / uh-seat-uh-MIN-oh-fin / KAF-feen (ecks-SAID-rin)

Most students remember **Excedrin Migraine** by the rationale for the combination of **aspirin with acetaminophen and caffeine** that is inflammation, analgesia, and vasoconstriction respectively.

A Prescription only NSAID Analgesic

Melox̲icam (brand, Mobic)
mel-OX-eh-kam (MO-bik)

The "-icam,"-i-c-a-m suffix in the generic name **meloxicam** lets you know it's an NSAID. A student used the "bic" in **Mobic** to remember it treats "big" swelling.

A COX-2 inhibitor Prescription only NSAID analgesic

Celecox̲ib (brand, Celebrex)
sell-eh-COCKS-ib (SELL-eh-breks)

The "–coxib, c-o-x-i-b" suffix lets you know celecoxib is a selective COX-2 inhibitor. This contrasts regular NSAIDs like **ibuprofen** and **naproxen** that inhibit both COX-1 and COX-2, causing the stomach distress so commonly caused by drugs in the NSAID class. **Celecoxib** causes less G-I irritation due to its lack of COX-1 inhibition.

The commercials for brand **Celebrex** talk about celeb̲rating relief from inflammatory conditions.

MUSCULOSKELETAL SECTION II. OPIOIDS AND NARCOTICS

Opioids relieve pain, but have addiction potential and can cause side effects like euphoria, sedation, constipation, and miosis that manifests as pinpoint pupils. A student remembered this by noticing that the words "opioid" and "miosis" both have two little dots over the two i's that look like – pinpoint pupils.

The Drug Enforcement Agency (D-E-A) categorizes the addictive potential of medications, including these opioids, using a drug scheduling system.

Schedule I drugs are illegal substances and have no medical value, such as **heroin**.

Schedule II drugs are potentially addicting, such as the individual drugs **fentanyl** or **morphine** and the combination products **hydrocodone with acetaminophen** or **oxycodone with acetaminophen.**

Schedule III drugs are less addicting and include **acetaminophen with codeine.**

Schedule IV drugs include some sedative-hypnotics (sleeping pills) such as **zolpidem** and mixed opioid analgesics like **tramadol**

Schedule V drugs are often cough medicines like **guaifenesin with codeine** that include codeine, but not **codeine** as a drug alone. Codeine as a drug alone is DEA schedule II.

FOUR SCHEDULE II OPIOID ANALGESICS

1) Morphine (brand names, Kadian or M-S Contin)
MORE-feen (KAY-dee-en, EM-ES KON-tin)

The generic name **morphine** comes from the ancient Greek god of dreams, Morpheus.

The brand **Kadian** might come for cir**cadian** (the twenty-four-hour cycle) because **Kadian** is an extended-release morphine formulation.

M-S Contin stands for **m**orphine **s**ulfate **contin**uous release, referencing the M-S from the beginnings of "morphine sulfate," and the "Contin" from the start of "continuous;" **M-S Contin**.

2) Fentanyl (brand names, Duragesic or Sublimaze)
FEN-ta-nil (dur-uh-GEE-zic, SUB-leh-maze)

Fentanyl is troubling because it's dosed in micrograms, not milligrams. When it was time to give our three-month-old preemie **fentanyl** after her pyloric valve stenosis surgery in the Pediatric Intensive Care Unit (PICU), I made sure to check the calculated dose.

Duragesic is a long **dura**tion anal**gesic** and comes in a patch that provides relief for 72 hours; our mnemonic for Duragesic takes the "dura" from a long "duration" and the "gesic" from the ending of "analgesic."

Sublimaze is an injectable form of **fentanyl**.

3) Hydrocodone with Acetaminophen (brand, Vicodin)
high-droe-CO-done / uh-seat-uh-MIN-no-fin (VIE-co-din)

Hydrocodone and **oxycodone** differ slightly in chemical structure. **Oxycodone** has one more oxygen atom than **hydrocodone**.

4) Oxycodone with acetaminophen gets its name from the first three letters of *oxy*gen.

Hydrocodone with acetaminophen has only one hydrogen and gets its name from the first five letters of *hydro*gen.

The "c-o-d-i-n" in **Vicodin** looks like **codeine**; just drop two e's from codeine to get "codin" in "Vicodin," so you can remember they're related.

Oxycodone with Acetaminophen (brand, Percocet)
ox-e-CO-done / uh-seat-uh-MIN-no-fin (PER-coe-set)

The "cet" (c-e-t) in **Percocet** comes from one British pronunciation of ac̲etaminophen. Some students use that "codone" and "codeine" look a little alike to remember the similarity, but this is not a true stem and **Percocet** doesn't contain codeine. Drug companies add **acetaminophen** as a mild analgesic.

OPIOID ANALGESIC - SCHEDULE III

Acetaminophen with Codeine (brand names, Tylenol with Codeine or Tylenol number 3)
uh-seat-uh-MIN-no-fin with CO-dean

(TIE-len-all with CO-dean, TIE-len-all NUM-ber three)

I am not sure why, but when you say the generic name of **Vicodin**, you say "**hydrocodone** *with* **acetaminophen**," but when you talk about **codeine,** the order is reversed. It's phrased as "**acetaminophen** *with* **codeine**"

Students seem to remember both because of the reverse and opposite order. The "#3" in **Tylenol #3** refers to the amount of codeine in combination. For example:

Tylenol #2 has 15 mg codeine with 300 mg acetaminophen

Tylenol #3 has 30 mg codeine with 300 mg acetaminophen

Tylenol #4 has 60 mg codeine with 300 mg acetaminophen

MIXED-OPIOID RECEPTOR ANALGESIC - SCHEDULE IV

Tram a̲dol (brand, Ultram)
TRA-muh-doll (ULL-tram)

Tramadol only weakly affects opioid receptors. For this reason, the D-E-A did not classify **tramadol** as a controlled substance until 2014. The "adol, a-d-o-l" stem indicates that it's a mixed opioid analgesic.

Many students think of "tram wreck" and "train wreck" as a way to remember that U*ltram* is used for pain.

AN OPIOID ANTAGONIST

Na̲loxone (brand, Narcan)
nah-LOX-own (NAR-can)

The "nal, n-a-l" stem in **naloxone** indicates it's an opioid receptor antagonist, but the brand name **Narcan** with its elements "Narc" and "an," also hints at a na̲rcotic a̲ntagonist.

Naloxone is an opioid receptor antagonist used in opioid overdose situations. Often it's in the L-E-A-N, lean, acronym of emergency medicines **l̲idocaine, e̲pinephrine, a̲tropine,** and **n̲aloxone**.

MUSCULOSKELETAL SECTION III. HEADACHES AND MIGRAINES

Pharmacology becomes frustrating for students because one drug can have many uses. When we memorize something, often, we want a single concrete place to file that memory.

We will keep aspirin, ibuprofen, and naproxen in the original file as numbers 14, 15, and 16, and just make a note for the new therapeutic indication – headache.

Common OTC drugs for headache and migraine include the NSAIDs like **aspirin, ibuprofen, naproxen,** and the combination **aspirin, acetaminophen and caffeine** in the brand **Excedrin Migraine** reviewed earlier.

But, severe migraines may require drugs in the $5\text{-}HT_1$ receptor agonist class, such as **ele*triptan*** and **suma*triptan*.** These work by activating receptors that reduce the swelling associated with migraines. We call them **triptans** because those two syllables are easier to say than "5-hydroxytryptamine receptor agonists."

CHAPTER 2: MUSCULOSKELETAL

It's important to contrast the terms **agonist** and **antagonist** as opposites.

An agonist is a drug that activates a receptor while an antagonist blocks the receptor.

One student thought of agonists as when she first stepped on an elliptical exercise machine and it awakened the screen. She thought of antagonist as her own excuses why she couldn't work out.

Another metaphor you will often see is that of dating. There are some funny videos on YouTube about agonism versus antagonism. The agonist is usually someone who wants to get a date and the antagonist gets in the way of the matchmaking.

WE'LL PAIR THE TRIPTANS TO BETTER REMEMBER THEM.

1) Ele<u>triptan</u> (brand, Relpax)
el-eh-TRIP-tan (rel-PACKS)

Remember **eletriptan** from its suffix "triptan, t-r-i-p-t-a-n" so you can recognize the many other medications in the triptan class.

The brand name **Relpax** combines "Rel" for "relief" and "pax," the Latin word for "peace." I often confused whether **Relpax** was an agonist or antagonist, but use the "agony" of a migraine to remember triptans as agonists.

2) Suma<u>triptan</u> (brand, Imitrex)
Sue-ma-TRIP-tan (IM-eh-treks)

Remember **sumatriptan** from its suffix "-triptan." Students say it "trips up" a headache.

Also, the "i-m" in the brand name **Imitrex** reminds you there is an intramuscular (I-M) form for patients who have such a severe migraine that they can't take anything by mouth.

Musculoskeletal Section IV. DMARDs and Rheumatoid Arthritis

DMARD stands for **D**isease-**M**odifying **A**nti-**R**heumatic **D**rug, which means it works against *rheumatoid arthritis*, an autoimmune disorder. These drugs reduce the progression of the disease as opposed to the treatment of *osteoarthritis*, a condition in which the body has worn down and the joints are inflamed.

Both conditions respond to NSAID anti-inflammatories such as **ibuprofen** or **naproxen**. Additionally, glucocorticoids, such as **prednisone,** can further help reduce inflammation in the joints.

Prescribers use a special class of drugs, the immune-suppressing drugs called DMARDs like **methotrexate**, **abatacept**, and **etanercept** for rheumatoid arthritis.

The Non-biologic DMARD

Methotrexate (brand, Rheumatrex)
meth-oh-TREKS-ate (ROOM-uh-treks)

The "trexate, t-r-e-x-a-t-e" stem helps remind you this is a DMARD. One student came up with "Meth o T-Rex ate the rheumatic inflammate;" **Methotrexate**.

The "rheuma" in the brand name **Rheumatrex** reminds you it relieves rheumatoid arthritis.

There are two biologic DMARDs we'll talk about.

1) Abatacept (brand, Orencia)
uh-BAT-uh-sep" (or-EN-see-uh)

Abatacept has a complex stem. The "-ta-", t-a, infix in **abatacept** means it's going after T-cell receptors and the suffix "–cept, c-e-p-t" means that it's a re**cept**or molecule, either native or modified.

2) Eta<u>ner</u>cept (brand, Enbrel)
eh-ta-NER-sept (EN-brell)

Learn both **etanercept** and **abatacept** from the suffix "–cept" (c-e-p-t), with the added sub-stem "-nercept" or "-tacept" respectively.

The "-ner-" (n-e-r) infix in **etanercept** points out that it goes after tumor <u>necr</u>osis factor receptors, where we're noting the letters n-e-r from "necrosis."

MUSCULOSKELETAL SECTION V. OSTEOPOROSIS

Don't confuse *osteoarthritis*, a joint disease, with *osteoporosis*, a thinning in bone tissue density. Drugs for osteoarthritis include the NSAIDs. A drug class for osteoporosis that builds bone back up includes the bisphosphonates.

Humans' bones grow slowly. Therefore, patients can take drugs like **alendronate** weekly, and **ibandronate** monthly. A special precaution for patients using **alendronate** is to not lie down for 30 minutes after taking the medication; for **ibandronate**, it's 60 minutes.

TWO BISPHOSPHONATES FOR OSTEOPOROSIS

1) Alen<u>dronate</u> (brand, Fosamax)
uh-LEN-dro-nate (FA-seh-max)

Memorize **alendronate** as a calcium metabolism regulator from its "dronate, d-r-o-n-a-t-e" stem. Students like to remember that "drone" rhymes with bone.

Many students have said that the brand name **Fosamax** looks like "fossil."

2) Iban<u>dronate</u> (brand, Boniva)
eh-BAND-row-nate" (bo-KNEE-vuh)

Again, the "dronate" suffix should be your key to the drug class, but having the first three letters of bone in the brand name **Boniva** helps. The generic **ibandronate** contains all the letters in **Boniva** except the "v."

MUSCULOSKELETAL SECTION VI. TWO MUSCLE RELAXANTS

The D-E-A doesn't schedule the muscle relaxant **cyclobenzaprine,** but does schedule the benzodiazepine **diazepam** as C-IV (Class four). Both drugs provide muscle relaxation and relief from muscle spasms.

Cyclobenzaprine (brand, Flexeril)
sigh-clo-BENDS-uh-preen (FLEX-er-ill)

Cyclobenzaprine helps you get bending again, using the 'b-e-n-z' in the center of the generic name. The brand *Flex*eril improves *flex*ibility.

Dia<u>zep</u>am (brand, Valium)
dye-AY-zeh-pam" (VAL-e-um)

Diazepam and benzo**diazep**ine, diazepam's drug class, have similar letters. Note the v-a-l in <u>Val</u>erian root, which is an herbal remedy for anxiety has the same three initial letters as **Valium**. Some think of **Valium** as relaxing both anxiety and muscles in the same way.

MUSCULOSKELETAL SECTION VII. GOUT

Gout is an inflammatory arthritis we can treat acutely (right away) with an NSAID like **ibuprofen**. We can also treat gout prophylactically (ahead of time) by reducing uric acid, a major component in the crystals that cause the gouty pain.

CHAPTER 2: MUSCULOSKELETAL

Drugs that alter uric acid levels include **allopurinol** and **febuxostat**.

Two Uric acid reducers

1) Allopurinol (brand, Zyloprim)
a-loe-PURE-in-all (ZY-low-prim)

Within **allopurinol**, you can see "uri, u-r-i," which corresponds to the uric acid the medication reduces. You can also remember this is an anti-arthritic by thinking of the joints as becoming "all-pure-and-all," mimicking the generic name, allopurinol.

2) Febuxostat (brand, Uloric)
fe-BUCKS-oh-stat (YOU-lore-ick)

The "xostat, x-o-s-t-a-t" stem in **febuxostat** indicates a xanthine oxidase inhibitor that prevents uric acid from forming. The "x" and "o" in the stem included in the generic name match these first letters in xanthine oxidase. The brand name **Uloric** looks like "U" "lower" "uric acid;" **Uloric**.

Chapter 2 Quiz: Musculoskeletal, Drug stem and Drug classification practice

I'm going to read a generic name for you and I want you to think about the stem, if any, and the class of medication.

1. Aspirin no stem, NSAID
2. Ibu<u>profen</u> profen, p-r-o-f-e-n, NSAID
3. Na<u>proxen</u> no stem, but proxen, p-r-o-x-e-n, looks like profen, p-r-o-f-e-n, NSAID
4. Acetaminophen no stem, non-narcotic analgesic
5. Aspirin with Acetaminophen and Caffeine no stem, NSAID with a non-narcotic analgesic and vasoconstrictor
6. Melox<u>icam</u> – icam, i-c-a-m, – NSAID

7. Cele<u>coxib</u> – coxib, c-o-x-i-b, – NSAID, COX-2 selective
8. Morphine – no stem, opioid analgesic
9. Fentanyl – no stem, opioid analgesic
10. Hydrocodone with acetaminophen – no stem, opioid analgesic and non-narcotic analgesic
11. Oxycodone with acetaminophen no stem, opioid analgesic and non-narcotic analgesic
12. Acetaminophen with codeine no stem, non-narcotic analgesic and opioid analgesic
13. Tram<u>adol</u> adol, a-d-o-l, mixed opioid receptor agonist and antagonist
14. <u>Nal</u>oxone - nal, n-a-l, opioid receptor antagonist
15. Ele<u>triptan</u> triptan, t-r-i-p-t-a-n, – 5-HT$_1$ receptor agonist for acute migraine
16. Suma<u>triptan</u> triptan, t-r-i-p-t-a-n, – 5-HT1 receptor agonist for acute migraine
17. Metho<u>trexate</u> trexate, t-r-e-x-a-t-e, – *non biologic* Disease modifying anti-rheumatic drug, DMARD
18. Aba<u>tacept</u> tacept, t-a-c-e-p-t, –*biologic* Disease modifying anti-rheumatic drug, DMARD
19. Eta<u>nercept</u> nercept, n-e-r-c-e-p-t, *biologic* Disease modifying anti-rheumatic drug, DMARD
20. Alen<u>dronate</u> dronate, d-r-o-n-a-t-e, - bisphosphonate for osteoporosis
21. Iban<u>dronate</u> dronate, d-r-o-n-a-t-e, - bisphosphonate for osteoporosis
22. Cyclobenzaprine no stem, muscle relaxer
23. Di<u>azepam</u> azepam, a-z-e-p-a-m, – benzodiazepine stem, classified in this chapter as a muscle relaxer
24. Allopurinol no stem, anti-gout prophylaxis
25. Febu<u>xostat</u> xostat, x-o-s-t-a-t, anti-gout prophylaxis

Musculoskeletal: Memorizing the Chapter

"M" Musculoskeletal

Three OTC NSAIDs, **aspirin, ibuprofen, naproxen**, and then **acetaminophen,** a non-narcotic analgesic. Combine **aspirin with acetaminophen** and **caffeine** to make **Excedrin Migraine.** Use migraine (M) and caffeine (C) to connect **meloxicam** (M) and **celecoxib** (C). Group opioids by DEA class. Start with DEA Schedule II **morphine**, the original agent; then **fentanyl, hydrocodone with acetaminophen** and **oxycodone with acetaminophen.** Next comes DEA schedule III, **acetaminophen with codeine**; to DEA Schedule IV **tramadol**; then to the opioid *narcotic antagonist* **naloxone (brand name, Narcan);** to migraine agony *agonists* for migraine headache with **eletriptan** and **sumatriptan**; down from the head to rheumatoid arthritis inside the joints with the DMARD **methotrexate**, a non-biologic; to two biologic DMARDs; **abatacept** to **etanercept**. From joint pain on to the brittle bones with **alendronate** and **ibandronate** for osteoporosis. Then move out to the muscles with the muscle relaxers **cyclobenzaprine** and **diazepam** all the way down to the big toe for the gout meds **allopurinol** and **febuxostat**.

Memorizing Pharmacology

Can you recite these in order?

Aspirin	Morphine	Metho<u>trexate</u>
Ibu<u>profen</u>	Fentanyl	Aba<u>tacept</u>
Naproxen		Etanercept
	Hydrocodone with	
Acetaminophen	acetaminophen	Alen<u>dronate</u>
	Oxycodone with	Ibandronate
Aspirin with	acetaminophen	
		Cyclobenzaprine
Acetaminophen	Acetaminophen	Di<u>azepam</u>
And Caffeine	with codeine	
		Allopurinol
Melox<u>icam</u>	Tram<u>adol</u>	Febuxostat
Cele<u>coxib</u>	N<u>al</u>oxone	
	Ele<u>triptan</u>	
	Suma<u>triptan</u>	

Once you're able to recite the list, you are ready to move on.

CHAPTER 3: RESPIRATORY

RESPIRATORY SECTION I. ANTIHISTAMINES AND DECONGESTANTS

The first thing we want to ask ourselves is "what is the difference between individual antihistamines?" There are many choices. It's useful to divide **antihistamines** into two generations: first and second.

Diphenhydramine is a *first*-generation antihistamine. In addition to its ability to improve allergy symptoms, it usually makes patients sleepy. The *second*-generation agents cannot pass through the blood-brain barrier and into the central nervous system, which limits the drowsiness patients experience. This second generation includes **cetirizine** and **loratadine**.

How do we remember the difference between the histamine- *one* receptor antagonists for allergy and the histamine-*two* receptor antagonists that reduce acid we heard in chapter 1?

My students came up with a helpful and creative visual to remember the difference between histamine-1 and histamine-2 receptors called the antihistamine snowman.

They pictured H_1 in the snowman's head because he has only one carrot for his nose, and allergies usually happen in the head or nose. They pictured H_2 as the snowman's belly because that's where gastroesophageal reflux disease (GERD) and hyperacidity happen.

However, patients often confuse the treatment for a runny nose (an antihistamine) with the treatment for a congested nose (a decongestant)

Nasal decongestants like **pseudoephedrine** constrict blood vessels in the nose and sinuses, reducing the formation of mucous. We can combine these *with* antihistamines for allergy symptoms, but decongestants and antihistamines are two separate drug classes.

When you see them together in one product, usually you will see the brand or generic name followed by a hyphen "D" for decongestant, like the brand **Claritin-D** to indicate that **pseudoephedrine** is a product ingredient. These pseudoephedrine products are not over-the-counter (OTC), but behind the counter (BTC) because you have to show an ID to purchase them.

Let's look at some specific antihistamines and decongestant products and how we can remember them.

AN OTC 1ST-GENERATION ANTIHISTAMINE

Diphenhydramine (brand, Benadryl)
dye-fen-HIGH-dra-mean (BEN-uh-drill)

Many students on YouTube videos and Quizlet notecards mistake **diphenhydramine's** "i-n-e" as a stem because many antihistamines end in "ine."

However, **morphine**, an opioid, ends in "een" and roughly 20% of *all* generic names do. There is not really a good stem for some generic antihistamine names because of this.

You can remember the brand name by recognizing **Benadryl** benefits you by drying up your runny nose; taking the "ben" from "benefits" and the "dry" from "drying."

Some students also associated the capital "B" in **Benadryl** with the BBB, the Blood Brain Barrier, which **Benadryl** *can* pass through.

Manufacturers use **diphenhydramine** as the "P-M" in many sleep aids, so associating the "B," "e," and "d" in **Benadryl** with bedtime also makes sense.

Two OTC 2nd-generation Antihistamines

1) Cetirizine (brand, Zyrtec)
seh-TIE-rah-zine (ZEER-teck)

I pronounce the "t-i-r" in **cetirizine** as tear, like teardrop, and I think of **cetirizine** protecting me from tearing, from eyes affected by allergies.

2) Lor<u>at</u>adine (brand, Claritin)
lore-AT-uh-dean (KLAR-eh-tin)

The "–atadine, a-t-a-d-i-n-e" stem (used to be "–tadine" t-a-d-i-n-e) helps distinguish **loratadine** from the "–tidine" (t-i-d-i-n-e) stem in the H₂ receptor antagonists **famotidine** and **ranitidine**.

The **Claritin** "clear" commercials resonate with the brand name drug's function of *clearing* one's head from allergies or *clear* eyes relieved from allergy.

A Combination 2nd-generation OTC Antihistamine and Decongestant

Lor<u>at</u>adine with Pseudoephe<u>drine</u> (brand, Claritin-D)
lore-AT-uh-dean / Sue-doe-uh-FED-rin (KLAR-eh-tin dee)

Because **pseudoephedrine** has 15 letters, manufacturers abbreviate it as "hyphen D" for decongestant. Therefore, adding **loratadine**, an antihistamine, to **pseudoephedrine** a decongestant, helps with both runny and stuffy noses.

Three Decongestants That Don't Require a Prescription

1) Pseudoephe<u>drine</u> (brand, Sudafed)
Sue-doe-uh-FED-rin (Sue-duh-FED)

If you take out the second "e" and drop the "rine" (r-i-n-e) from **pseudoephedrine**, you get the pronunciation of **Sudafed**. One student said she was p-h-e-d up, "phed up," with being congested, and that's how she remembered it.

2) Phenylephrine (brand, NeoSynephrine)
FEN-ill-EF-rin (KNEE-oh-sin-EF-rin)

Phenylephrine sounds a bit like **pseudoephedrine** – they are both decongestants. Patients recognize *pseudoephedrine* by the "hyphen D" and *phenylephrine* by the "P-E" abbreviation in many cold preparations.

Phenylephrine is not as strong as **pseudoephedrine** and that's one reason it's available OTC, while **pseudoephedrine** is not.

3) Oxymetazoline (brand, Afrin)
ox-EE-meh-taz-oh-lin (AF-rin)

Oxymetazoline takes us from physically behind-the-counter (B-T-C) oral **pseudoephedrine** to *over*-the-counter nasal or oral **phenylephrine** to the *intranasal* decongestant **oxymetazoline**.

The "**Afrin**" brand name sounds a little like the "ephrine" that forms the ending of **phenyl*ephrine*** so you can relate the two as decongestants.

RESPIRATORY SECTION II. ALLERGIC RHINITIS STEROID, ANTITUSSIVES AND MUCOLYTICS

Let's start with pathophysiology. What is allergic rhinitis? How do we treat it?

Allergic rhinitis is an inflammation (*-itis*) of the nose (*rhin-*). We treat it with a local nasal steroid like **triamcinolone**. Nasal steroids don't work right away like a topical decongestant such as **oxymetazoline** might; rather, it takes weeks until a patient feels relief.

That's a summary of allergic rhinitis, what about the difference between cough and chest congestion?

Because the brand name "**Robitussin**" has been associated with cough relief so long, many students mistakenly confuse plain **Robitussin** (just **guaifenesin**) with **Robitussin DM (guaifenesin and dextromethorphan)**. In **Robitussin DM**, the **guaifenesin** acts as a mucolytic to lyse (break up) mucous and chest congestion, while the **D-M (dextromethorphan)** acts as the antitussive or cough suppressant. In severe cases, we employ a **codeine**-based prescription product such as **guaifenesin with codeine.**

Let's take a look at these specific products.

OTC Allergic rhinitis steroid nasal spray

Triamcinolone (brand, Nasacort Allergy twenty four hour)
try-am-SIN-oh-lone (NAY-zuh-cort)

There is no recognized stem in **triamcinolone**, but often "o-n-e," pronounced like I "own" something, matches the "o-n-e" at the end of **testosterone**, a more familiar steroid.

The brand name **Nasacort** Allergy twenty-four hour reads like a story: "Nasa" for nose, "cort" for corticosteroid, "Allergy" for allergic rhinitis, and "twenty four hour" for how long it works; Nasacort Allergy twenty four-hour.

OTC Mucolytic with Antitussive

Guaifenesin with Dextromethorphan
(brand names, Mucinex DM and Robitussin DM)
*gwhy-FEN-uh-sin / decks-trow-meth-OR-fan
(MEW-sin-ex dee-em, row-beh-TUSS-in dee-em)*

Memorizing Pharmacology

Guaifenesin, pronounced as if you put a "g" in front of "why," is a mucolytic, something that lyses or breaks up mucous. Students remember this because Mr. Mucus from the **Mucinex**-brand commercials is green, and green also starts with a "g."

Robitussin "robs" your cough and "tussin" resembles "tus-sive." Anti*tussives* are anti-cough medicines.

PRESCRIPTION-ONLY MUCOLYTIC WITH NARCOTIC ANTITUSSIVE

Guaifenesin with Codeine (brand, Cheratussin A-C)
gwhy-FEN-uh-sin / CO-dean (CHAIR-uh-tuss-in ay-see)

Sometimes **dextromethorphan** (**DM**) isn't enough and the patient needs **codeine** to suppress a cough. Most students know **codeine**, but cough and **codeine** both start with "c-o" and that seems to help.

The "chera" in the brand **Cheratussin** comes from the product's cherry flavoring. Some are not sure what the "AC" means; some students think "anti-cough" although it's probably "and codeine."

RESPIRATORY SECTION III. ASTHMA

An initial question to ask here is, "What are the two components of asthma we're trying to affect?"

Asthma is a disease of bronchoconstriction (the lung's branches tighten) and inflammation. For immediate relief *during* an attack, an **albuterol** inhaler is the short-acting bronchodilator that will reverse the bronchoconstriction. Oral steroids such as **methylprednisolone** and **prednisone** help reduce lung inflammation *after* a severe attack.

The combination inhalers **fluticasone with salmeterol** or **budesonide with formoterol** provide relief from both inflammation and bronchoconstriction by combining a steroid (for inflammation) and long-acting beta two receptor agonist (which bronchodilates). Notice that many steroids have "sone" (s-o-n-e) which is *not* an official stem, at the end of their names, and that beta 2 receptor agonists have the stem "terol, t-e-r-o-l" in the suffix. These long-acting combinations prophylactically prevent asthma attacks.

Besides bronchodilating beta 2 receptor agonists and anti-inflammatories, what other drug classes work to treat asthma?

1) **Ipratropium** and **tiotropium** are anticholinergic medications. Often these medications cause dry mouth, constipation and other unwanted adverse effects. However, **ipratropium** and **tiotropium** affect the smooth muscle of the lungs, allowing for bronchodilation and relaxation of the bronchi.

Both **albuterol** and **ipratropium** bronchodilate, they just work differently. **Albuterol** is an *agonist* of beta 2 receptors and **ipratropium** is an *antagonist* of acetylcholine (A-C-h).

2) **Montelukast** inhibits leukotriene receptors. Leukotrienes cause bronchoconstriction, a process that protects the lungs against foreign contaminants. Drugs in this class end in "–lukast, l-u-k-a-s-t"

3) **Omalizumab** is an I-g-E antagonist and biologic monoclonal antibody.

Let's put this together with more detail and mnemonics.

Two Oral steroids

1) Methyl<u>pred</u>nisolone (brand, Medrol)
meth-ill-pred-NISS-uh-lone (MED-rol)

A student connected "pred" and "predator of inflammation" for generic **methylprednisolone**. **Methylprednisolone** comes in a 6-day, 21-pill dose pack that gives patients 6 tablets on the first day, 5 on the 2nd, 4 on the 3rd, 3 on the 4th, 2 on the 5th, and 1 on the 6th. This dosing reminds us to *taper* steroids to allow the adrenal glands time to resume normal function.

2) Prednisone (brand, Deltasone)
PRED-ni-sewn (DEALT-uh-sewn)

Most students know **prednisone**, but many steroid compounds have this unofficial "sone" (s-o-n-e) ending. In this case, the "pred, p-r-e-d" is the official stem in the *prefix* position. Note, we can also find this stem in the middle of the drug name as a meaningful *infix* in the steroid methyl**pred**nisolone.

TWO INHALED STEROID WITH LONG-ACTING BETA$_2$-RECEPTOR AGONIST COMBINATIONS

1) Budesonide with Formoterol (brand, Symbicort)
byou-DES-uh-nide / four-MOE-ter-all (SIM-buh-court)

Similar to the **fluticasone** paired with **salmeterol**, **budesonide** has the "sone, s-o-n-e" syllable in the middle of its name, and is pronounced "sone" even though it's spelled "s-o-n."

The "terol, t-e-r-o-l" stem in **formoterol** indicates a beta 2 agonist bronchodilator. You can think of the "S-y-m" in the brand name **Symbicort** as symbiotic, meaning "working with," plus "c-o-r-t" for corticosteroid: Sym-bi-cort.

2) Salmeterol with Fluticasone (brand, Advair)
flue-TIC-uh-sewn / Sal-ME-ter-all (ADD-vair)

Recognize the steroid **fluticasone** by the unofficial "sone," and the long-acting bronchodilator **salmeterol** by the "terol" stem. The brand name **Advair** seems like "<u>ad</u>d two drugs to get <u>air</u>." Ad-vair.

An Inhaled Steroid

Fluticasone (brand names, Flovent HFA, Flovent Diskus, and Flonase)
flue-TIC-uh-sewn (FLOW-vent)

The "sone, s-o-n-e" ending, while not an official stem, is a useful clue that **flutica*sone*** is a steroid. In the past, inhalers felt cold when patients used them because the chlorofluorocarbon (C-F-C) propellant spray was similar to Freon, the refrigerant used in air conditioning. However, C-F-Cs damage the ozone layer, so the new ozone-safe <u>h</u>ydr<u>o</u>fluoro<u>a</u>lkane (H-F-A) replaces the CFC propellant.

The brand name **Flovent H-F-A** cleverly uses the first two letters of the generic name **fluticasone**, incorporates f-l-o from "airflow" and adds, "vent" to let the patient know this is administered in the mouth.

The diskus, d-i-s-k-u-s inhaler is a device that looks like an Olympic discus, d-i-s-c-u-s. It employs a dry powder for inhalation rather than a propellant and liquid.

Flonase is the brand name for **fluticasone** administered nasally, and is available OTC.

A Beta 2 receptor agonist short-acting

Albu<u>terol</u> (brand, ProAir HFA)
Al-BYOU-ter-all (PRO-air aitch-ef-ay)

Albuterol's "terol, t-e-r-o-l" stem indicates it's a beta-2 adrenergic agonist that causes bronchodilation.

Note, albuterol's "terol" stem *does not* help you know if it's long acting *or* short-acting; that distinction must be memorized.

The brand name **ProAir HFA** is straightforward with "Pro" as in "I'm for it" and "Air" for airway.

BETA$_2$ RECEPTOR AGONIST WITH SHORT-ACTING ANTICHOLINERGIC

Albuterol with Ipratropium (brand, DuoNeb)
Al-BYOU-ter-al / Ih-pra-TROPE-e-um (DUE-oh-neb)

A "-terol" short-acting bronchodilator like **albuterol** can combine with a shorter acting anticholinergic like *ipratropium* (as compared to longer-acting *tiotropium*) in nebulized form to treat asthma symptoms faster.

Instructors use **atropine** as the prototype drug for the anticholinergics and you can see the "trop, t-r-o-p" stem in **atropine, ipratropium**, and **tiotropium**.

The brand name **DuoNeb** indicates a duo of drugs in nebulized form; Duo-Neb.

ASTHMA AND CHRONIC OBSTRUCTIVE PULMONARY DISEASE (C-O-P-D) MEDICATION - LONG-ACTING ANTICHOLINERGIC

Tiotropium (brand, Spiriva)
tie-oh-TROW-pee-um (Spur-EE-va)

Tiotropium has the same "trop, t-r-o-p" stem as **ipratropium** and is the long-acting version. As with the "terols," the beta$_2$ receptor agonists, you have to memorize which anticholinergic is long-acting versus short-acting. The brand name **Spiriva** takes the "spir, s-p-i-r" from respire, like respiration.

Just a note about the anticholinergic stems, the trop*ine* and trop*ium* indicate a difference in the number of attached atoms to the nitrogen atom. A "tropine" is a tertiary amine

CHAPTER 3: RESPIRATORY

with three attached atoms while the "tropium" is a quaternary amine with four. I would just go ahead and memorize either the whole tropine or tropium stem respectively.

ASTHMA MEDICATION - LEUKOTRIENE RECEPTOR ANTAGONIST

Monte<u>lukast</u> (brand, Singulair)
Mon-tee-LUKE-ast (SING-you-lair)

Leukotrienes that form in leukocytes (white blood cells) can cause inflammation. By blocking them, **montelukast** helps with the inflammatory component of asthma. The "-lukast, l-u-k-a-s-t" stem is similar to *leu*kotriene.

The **Singulair** brand name comes from its once daily "single" dosing and the "air" it helps to bring into the asthmatic lungs; **Singulair**.

ASTHMA - ANTI-I-G-E ANTIBODY

Oma<u>lizumab</u> (brand, Xolair)
oh-mah-liz-YOU-mab (ZOHL-air)

Like **infliximab** for ulcerative colitis, **omalizumab** is a biologic.

The "inf-" (i-n-f) is a prefix that separates it from other similar drugs.

The "-li-" (l-i) stands for immunomodulator (the target), the "-zu-" (z-u) stands for humanized (the source) and the "-mab" (m-a-b) is for <u>m</u>onoclonal <u>a</u>nti<u>b</u>ody.

There is a black box warning (a severe warning immediately at the beginning of the package insert) for the possibility of anaphylaxis after the first dose and even a year after the onset of treatment.

Therefore, health providers inject **omalizumab** where a medicine for treating anaphylaxis is available.

39

The brand name **Xolair** seems like, "exhale" and "air."

RESPIRATORY SECTION IV. ANAPHYLAXIS

Anaphylaxis is a special type of allergic overreaction of the body to something like an insect bite or bee sting. An e**pinephrine** injection quickly reverses the reaction, keeping the airway open.

Epinephrine (brand, EpiPen)
eh-peh-NEF-rin (EP-ee-pen)

The word **epinephrine** has a Greek origin. "Epi" means "above," and "neph" means "kidney." Above the kidney is the adrenal gland responsible for the body's natural release of **epinephrine**.

The Latinized version of **epinephrine** is adrenaline

The "ad" means "above" and "renal" means "kidney"; adrenaline.

Either version stimulates the fight or flight response that can help a patient in an anaphylactic state.

The **EpiPen** brand name comes from the injector device that looks somewhat like a pen.

CHAPTER 3 RESPIRATORY, DRUG STEM AND DRUG CLASSIFICATION PRACTICE

I'm going to read a generic name for you and I want you to think about the stem, if any, and the class of medication.

1. **Diphenhydramine** no stem, 1st generation antihistamine
2. **Cetirizine** no stem, 2nd generation antihistamine
3. **Lor<u>atadine</u>** atadine, a-t-a-d-i-n-e, 2nd generation antihistamine

4. **Lor<u>atadine</u>-D** atadine, a-t-a-d-i-n-e, 2nd generation antihistamine and decongestant pseudoephedrine, drine, d-r-i-n-e
5. **Pseudoephe<u>drine</u>** drine, d-r-i-n-e, decongestant
6. **Phenylephrine** no stem, nasal or liquid decongestant
7. **Oxymetazoline** no stem, nasal decongestant
8. **Triamcinolone no** stem, but some use the "o-n-e" pronounced "own" to remember it's a nasal steroid
9. **Guaifenesin with D-M** (dextometh<u>orphan</u>) orphan, o-r-p-h-a-n, OTC mucolytic with antitussive
10. **Guaifenesin with codeine** no stem, prescription only mucolytic with opioid antitussive
11. **Methyl<u>pred</u>nisolone** pred, p-r-e-d, steroid
12. **<u>Pred</u>nisone** pred, p-r-e-d, steroid
13. **Budesonide with Formo<u>terol</u>** use "sewn," s-o-n, with caution, not an official stem, steroid, with terol, t-e-r-o-l, beta 2 agonist bronchodilator
14. **Fluticasone with Salme<u>terol</u>** use "sewn," s-o-n-e, with caution, not an official stem, steroid, with terol, t-e-r-o-l, beta 2 agonist bronchodilator
15. **Fluticasone**, I sound like a broken record, but use "sewn," s-o-n-e, with caution, not an official stem, steroid
16. **Albu<u>terol</u>** terol, t-e-r-o-l, beta 2 agonist bronchodilator
17. **Albu<u>terol</u> with Ipra<u>tropium</u>** terol, t-e-r-o-l in albuterol is a beta 2 agonist bronchodilator and tropium, t-r-o-p-i-u-m is a short-acting anticholinergic bronchodilator in ipratropium
18. **Tio<u>tropium</u>** tropium, t-r-o-p-i-u-m, long-acting anticholinergic bronchodilator
19. **Monte<u>lukast</u>** lukast, l-u-k-a-s-t, leukotriene receptor antagonist
20. **Oma<u>lizumab</u>** lizumab, l-i-z-u-m-a-b, monoclonal antibody for asthma
21. **Epinephrine** no stem, anaphylaxis injection

Respiratory: Memorizing the Chapter

"R" Respiratory

Start with the 1st-generation O-T-C antihistamine **diphenhydramine**, and then go to the 2nd-generation **cetirizine** (which starts with a C) then alphabetically to 2nd-generation **loratadine** (with an L). Add a decongestant to make **loratadine-D** behind-the-counter (B-T-C), and then move to what the "hyphen D" stands for – **pseudoephedrine**, the decongestant alone. Then walk into the OTC aisle to get the oral or nasal decongestant **phenylephrine** and move up to the nose with the intranasal-only decongestant **oxymetazoline**. Stay in the nose with an OTC intranasal glucocorticoid **triamcinolone**. Move from *nasal* congestion to *chest* congestion with **guaifenesin with dextromethorphan** to another antitussive combination behind the prescription counter for prescription-only, **guaifenesin with codeine**. If that doesn't work and the coughing inflames your lungs, you might need an oral steroid like **methylprednisolone** (which starts with an M and has "pred" p-r-e-d in the middle) or **prednisone** (which follows alphabetically, beginning with P). After the acute attack, you might find you have asthma and need to use a prophylactic inhaled steroid in combination with a long-acting beta$_2$ receptor agonist, either **budesonide** with **formoterol** or **salmeterol** with **fluticasone**. You could individually use the steroid **fluticasone** or **albuterol**, the short-acting beta 2 receptor agonist. The beta 2 receptor agonist **albuterol** combined with anticholinergic **ipratropium** makes the duo in the brand, **DuoNeb**, or alternatively, the long-acting anticholinergic **tiotropium** can be given alone. If that doesn't work, use **montelukast** against leukotrienes or **omalizumab** against IgE, but remember that **omalizumab** has a black box warning about anaphylaxis, which might necessitate an injection of **epinephrine**.

CHAPTER 3: RESPIRATORY

Can you recite these in order?:

Diphenhydramine	Triamcinolone	
		Albu<u>terol</u>alone
Cetirizine	Guaifenesin with	Albu<u>terol</u> with
Loratadine	dextromethorphan	Ipra<u>tropium</u>
	Guaifenesin with	Tiotropium
Lor<u>atadine</u>-D	codeine	
Pseudoephe<u>drine</u>		Monte<u>lukast</u>
	Methyl<u>pred</u>nisolone	
Phenylephrine	<u>Pred</u>nisone	Oma<u>lizumab</u>
Oxymetazoline		
	Budesonide with	Epinephrine
	Formo<u>terol</u>	
	Fluticasone with	
	Salme<u>terol</u>	
	Fluticasone alone	

Once you're able to recite the list, you are ready to move on.

CHAPTER 4: IMMUNE

IMMUNE SECTION I. OTC Antimicrobials

There are many, many antimicrobial medications, how do we remember them all? Just like the previous chapters, we want to break them down into reasonable partitions.

The word "antimicrobials" means "against microbes." We generally divide them into three major classifications: **antibiotics** (drugs for bacteria), **antifungals** (drugs for mycoses or fungi), and **antivirals** (drugs for viruses).

Antibiotic brand names don't give good information about drugs because they often derive from generic names. As such, creating linkages between drugs with the same mechanism of action, so you can group similar antibiotics, becomes critical.

For example, penicillins and cephalosporins, along with **vancomycin**, affect bacterial cell walls. By putting them near each other on this ordered list of 200, you can group them into a larger category.

Antifungals and antivirals, in contrast, have excellent brand names that allude to their therapeutic effect. Most students try to be ultra-efficient and *only* memorize generic names.

Just as you have more information if you know a person's first and last name, you know more about a drug by memorizing both generic and brand names. This backup information is critical under the stress of exams or clinical practice.

I've grouped four over-the-counter medications into the three main categories: antibiotic, antifungal, antiviral to set up the same framework in the prescription-only section.

Let's start with an OTC antibiotic cream you're probably familiar with. You might not recognize **Neomycin with Polymyxin B** and **Bacitracin,** but you *do* recognize the brand name **Neosporin.**

knee-oh-MY-sin / pall-EE-mix-en / bah-seh-TRACE-in (KNEE-oh-spore-in)

The **neomycin** component is an aminoglycoside generally toxic to the kidney (nephrotoxic) and ears (ototoxic) when used systemically (in the body). However, patients safely use *topical* preparations containing **neomycin** such as over-the-counter **Neosporin**.

The brand **Neosporin** takes "N-e-o and 's'" from **neomycin sulfate**, "p-o" from **polymyxin B**, and "r-i-n" from **bacitracin.**"

SECOND, LET'S LOOK AT AN OTC ANTIFUNGAL CREAM

Butenafine (brand, Lotrimin Ultra)
BYOO-ten-uh-feen (LOW-treh-min)

Butenafine treats topical fungal infections like ringworm, jock itch, and athlete's foot. Sometimes you will see the Latin names for these conditions: *tinea coporis* (ringworm), *tinea cruris* (jock itch), and *tinea pedis* (athlete's foot) respectively.

THIRD, LET'S LOOK AT TWO TYPES OF ANTIVIRAL, ONE ANTIVIRAL FOR PROPHYLAXIS (PREVENTING INFECTION) AND ONE ANTIVIRAL TO TREAT AN ACTIVE INFECTION.

Influenza Vaccine (brand names Fluzone, and Flumist)
in-FLU-en-zah VACK-seen (FLEW-zone, FLEW-mist)

While some children might need a prescription for the prophylactic **influenza vaccine**, most adults can walk up to the pharmacy counter and get a flu shot.

Be careful; generic names with "f-l-u," for example, **flucon-azole,** an antifungal, and **fluoxetine**, an antidepressant, refer to a fluorine atom in their chemical structure, not to the influenza virus.

The "flu" in the vaccine brand names **Fluzone** and **Flumist** indicates in<u>flu</u>enza, just like the brand **Tamiflu**, an oral medication for influenza infection that contains the syllable "flu."

The *nasal* vaccine **Flumist** provides an alternative for patients who don't want an injection.

Either way, knowing if a drug is a brand or generic name becomes critical to knowing the meaning of f-l-u.

OTC Antiviral (Acute infection)

Docosanol (brand, Abreva)
Do-cah-SAN-all (uh-BREE-vah)

Docosanol is a topical antiviral for cold sores caught and treated early. I thought, "who would pay twenty dollars for a small tube like that?" Then I thought of homecoming dances and ruined pictures with a large cold sore.

So, my mnemonic became "use **docosanol**, so you can go to the ball."

Abreva, the brand name, hints at therapeutic effect as **Abreva** "<u>abbrev</u>iates" the time a cold sore lasts.

IMMUNE SECTION II. Antibiotics
AFFECTING CELL WALLS

Bacteria have cell walls. Human cells don't (although they do have cell membranes). This introduces *selectivity*. If a drug targets a tissue or structure that bacteria have but humans don't, it should be *selective* for the bacteria and safe for the patient.

Penicillin's mechanism of action is to open a bacterium's cell wall, like popping a bubble, to kill it. This killing action is termed bactericidal.

However, sometimes we see resistance to a single antibiotic like **amoxicillin.** For example, a child with an ear infection finishes a week-long course of "the pink stuff" and remains sick.

We had this exact issue when our daughter had an adenoidectomy and tonsillectomy. She woke up in the middle of the night complaining of ear pain, even though she had an active amoxicillin prescription. We verified she had an ear infection and asked, "What's our next course of action?"

Amoxicillin with clavulanate adds the compound clavulanate to protect the **amoxicillin** against an enzyme bacteria produce called a beta-lactamase.

The enzyme acquired its name from the chemical structure (a beta lactam) that's in all penicillins. This additional component, **clavulanate**, augments, as its name brand says, **amoxicillin** allowing it to work in cases where it had previously failed.

That was one therapeutic option our physician gave us. The other option was to give a 3rd generation cephalosporin antibiotic.

The **cephalosporin** antibiotics can have cross-sensitivity with penicillins. Patients allergic to one may be allergic to the other, but this is quite rare. We classify cephalosporins into generations.

The first-generation drugs, such as **cephalexin** don't penetrate the cerebrospinal fluid (CSF), have poor gram-negative bacterial coverage (gram-negative bacteria have an extra protective layer and do not take up a gram stain), and are subject to deactivation by beta-lactamase producing bacteria. This 1st generation cephalosporin would not help our daughter.

However, as we move to third-generation **ceftriaxone** and fourth-generation **cefepime,** we get good penetration into the CSF, good gram-negative coverage and the antibiotics cover bacteria resistant to beta-lactam drugs.

So, back to our story. The ultimate choice for our daughter was to give a 3rd generation oral cephalosporin, called **cefdinir**, brand name **Omnicef**, to treat this beta-lactamase resistant infection.

Another antibiotic that affects bacteria cell walls as its mechanism of action is **vancomycin. Vancomycin** is at times the last line of defense against a sometimes-deadly bacterial infection like methicillin-resistant *Staphylococcus aureus* (MRSA) or M-R-S-A.

To minimize resistance, a special protocol dictates who can and cannot get **vancomycin.**

Vancomycin has special dosing requirements for patient safety and often pharmacists use their expertise to dose it appropriately so that patients receive optimal drug therapy.

In rare cases, **vancomycin** can cause a hypersensitivity reaction called red man syndrome.

That was a lot of information to take in. Let's look more closely at the drug names that have established stems.

ANTIBIOTICS: PENICILLINS

Amox<u>icillin</u> (brand, Amoxil)
uh-mocks-eh-SILL-in (uh-MOCKS-ill)

Amoxicillin has the "cillin, c-i-l-l-i-n" stem that indicates its relationship to the penicillin family.

Memorizing Pharmacology

The "a-m-o" probably came from the fact that it's an "<u>amino</u>" penicillin.

The "cillin" stem sounds like "*cell*-in" and can help you remember that **amoxicillin** or, more generally, **penicillins** destroy the *cell* wall.

The brand name **Amoxil** simply removes an "i-c" and "l-i-n" from the generic name, amoxicillin, to make Amoxil.

A PENICILLIN CLASS ANTIBIOTIC WITH A BETA-LACTAMASE INHIBITOR

Amoxi<u>cillin</u> with clavulanate (brand, Augmentin)
uh-mocks-eh-SILL-in / clav-you-LAN-ate (awg-MENT-in)

When **amoxicillin** alone doesn't work, we can think of the brand **Augmentin** as it <u>augment</u>s amoxi<u>cillin</u>'s defenses against the bacterial beta-lactamase enzyme with **clavulanate**.

I think of the "<u>clav</u>icle," the bone in your shoulder, as protective of the upper lung and associate **clavulanate** with that same protective effect as both start with c-l-a-v.

1ST GENERATION CEPHALOSPORIN ANTIBIOTIC

<u>Ceph</u>alexin (brand, Keflex)
sef-uh-LEX-in (KEH-flecks, not KEY-flecks)

With cephalosporins, a newer generation has better properties than the last, relative to what the prescriber is treating. Those advantages include better penetration into the cerebrospinal fluid (C-S-F), better gram- negative coverage, and better resistance to beta-lactamases.

You may lose some gram-positive coverage as you move up the spectrum, however. The "ceph, c-e-p-h" is an old stem from the first generation. The new stem "cef, c-e-f) identifies the newer generations.

CHAPTER 4: IMMUNE

That's how I remember **cephalexin** as first generation.

The brand name **Keflex, which begins with a "K,"** takes some letters from **cephal<u>ex</u>in** to make its name.

3RD GENERATION CEPHALOSPORIN ANTIBIOTIC

<u>Cef</u>triaxone (brand, Rocephin)
sef-try-AX-own (row-SEF-in)

In the generic name, **ceftriaxone's** "cef-" (c-e-f) indicates it's a cephalosporin.

There is a "tri" (t-r-i) in the generic name that you can use to remember it is 3rd generation.

Rocephin, the brand name, seems to come from Hoffman-La<u>R</u>oche's patent.

The drug company took the "Ro" (r-o) from "La<u>Ro</u>che," and the "ceph" (c-e-p-h) and "in" from <u>ceph</u>alospor<u>in</u> to make **Ro-ceph-in**.

4TH GENERATION CEPHALOSPORIN ANTIBIOTIC

<u>Cef</u>epime (brand, Maxipime)
SEF-eh-peem (MAX-eh-peem)

Cefepime is a fourth-generation cephalosporin. I've remembered it by thinking of four letters "p-i-m-e" that are in both the brand name **Maxipime** and the generic name **cefepime**.

Also, at the time, **Maxipime** was the *max*imum generation, the fourth and highest.

However now there is a fifth-generation cephalosporin, but that old mnemonic stayed in my brain.

Glycopeptide antibiotic

Vancomycin (brand, Vancocin)
van-co-MY-sin (VAN-co-sin)

Vancomycin's "mycin, m-y-c-i-n" stem isn't very useful for finding its therapeutic class.

All "mycin" really means is that chemists derived **vancomycin** from the *Streptomyces* bacteria, taking the m-y-c from *Streptomyces*.

I remember the function as "**vancomycin** will vanquish MRSA."

To remember the brand name **Vancocin**, remove the "my" from **vancomycin,** to get V-a-n-c-o-c-i-n. **Vancocin**.

Immune Section III. Antibiotics – Protein synthesis inhibitors – Bacteriostatic

Tetracycline antibiotics

We name bacteriostatic **tetracyclines** like **doxycycline** and **minocycline** after the four (tetra) member chemical ring (cycline). Tetracyclines and fluoroquinolones both cause photosensitivity and chelation (binding with cations such as the Ca++ in milk or antacids).

Macrolide antibiotics

We sometimes call **macrolides** "erythromycins" after one of the original drugs in the class. Patients take **azithromycin** as a double dose on the first day and a single dose the following four days. The double dose is called a *loading dose*.

This once-daily dosing improves patient compliance.

Patients take **clarithromycin** twice a day – note there is a "bi" prefix in clarithromycin's brand name, **Biaxin**, to help you remember.

We dose **erythromycin** four times a day and this is probably why it has fallen out of favor. Too many daily doses.

Lincosamide antibiotic

Dentists use **clindamycin** for dental prophylaxis when a patient is penicillin allergic.

Patients use it topically for severe acne.

When used orally, it can cause a severe condition known as pseudomembranous colitis, also known as antibiotic-associated diarrhea (AAD).

Oxazolidinone antibiotic

Linezolid is an oxazolidinone antibiotic that can work on both **methicillin**-resistant *Staphylococcus aureus* (M-R-S-A) and **vancomycin**-resistant enterococci (V-R-E).

Let's look more closely at the two tetracycline antibiotics

1) Doxy<u>cycline</u> (brand, Doryx)
docks-ee-SIGH-clean (DOOR-icks).

I use the "d" in **doxycycline** to remind me that dentists use it to treat periodontal disease.

Doryx, the brand name, takes the first four letters of **doxycycline** and adds an "r," scrambling them a bit to create Doryx.

2) Mino<u>cycline</u> (brand, Minocin)
MIN-oh-SIGH-clean (MIN-oh-sin)

Minocycline, like **doxycycline** has the **tetracycline** class "cycline, c-y-c-l-i-n-e" stem. To create the brand name **Minocin**, the manufacturer dropped the "c-y-c-l" and last "e," from minocycline to get M-i-n-o-c-i-n, Minocin.

THREE MACROLIDE ANTIBIOTICS

1) Azi<u>thromycin</u> (brand names Zithromax or Z-Pak)
ay-zith-row-MY-sin" (ZITH-row-max)

To recognize the three macrolides, **azithromycin, clarithromycin**, and **erythromycin**, you will see the "mycin, m-y-c-i-n" ending, but also a possible infix of "thro, t-h-r-o."

Be careful: there *are* macrolides without this infix and stem.

The brand name **Zithromax** takes seven letters from **a**zi**thromy**cin to construct its name.

The **Zithromax Z-pak** is a convenient six-tablet package that includes a five-day course, two tablets for a loading dose on day one and one tablet for each thereafter.

2) Clari<u>thromycin</u> (brand, Biaxin)
Claire-ITH-row-my-sin (bi-AX-in)

Again, you can use the "thromycin" to help clue you in to this drug's class. Gastroenterologists prescribe **clarithromycin** for peptic ulcer disease (PUD) triple therapy along with **amoxicillin** and a proton pump inhibitor like **omeprazole**.

The "b-i" in the brand name **Biaxin** indicates the twice daily dosing from the Latin abbreviation b.i.d. for *bis in die*; twice a day.

3) Ery<u>thromycin</u> (brand, E-Mycin)
err-ith-row-MY-sin (E-MY-sin)

Some **erythromycin** tablets are bright red and that might be where it got its name. An erythrocyte is a red blood cell, and the word comes from connecting "erythro," the Greek for "red," and "cyte," for cell.

The brand name **E-mycin** comes from taking the "rythro" out of the generic name **erythromycin** to render **E-mycin**.

A Lincosamide Antibiotic

Clinda<u>mycin</u> (brand, Cleocin)
clin-duh-MY-sin (KLEE-oh-sin)

Most students remember the adverse effect C-DAD (<u>C</u>lostridium <u>d</u>ifficile-<u>A</u>ssociated <u>D</u>isease) because there is a "c" and a "da" right after each other in the generic name **clin<u>da</u>mycin**.

To make the brand name **Cleocin**, the manufacturer replaced the "i-n-d-a-m-y" in **clindamycin** with "e-o," for Cleocin.

An Oxazolidinone Antibiotic

Line<u>zolid</u> (brand, Zyvox)
LYNN-ez-oh-lid (ZIE-vocks)

The "zolid, z-o-l-i-d" stem in **linezolid** comes from the **oxazolidinone** class.

I think it's more helpful to think, "Man, **Zyvox** is zolid (solid); it treats two very difficult-to-kill organisms, Methicillin resistant staph aureus, M-R-S-A and vancomycin resistant enterococci, V-R-E."

IMMUNE SECTION IV. ANTIBIOTICS - PROTEIN SYNTHESIS INHIBITORS - BACTERICIDAL

I think of the "side" (s-i-d-e) in **aminoglyco<u>side</u>** and "cide, c-i-d-e" as in "cidal" to remind me these drugs are killers.

Bacteri*cidal* **aminoglycosides** can damage the kidneys (called nephrotoxicity) and ears (called ototoxicity).

Two Aminoglycoside antibiotics

1) Ami**kacin** (brand, Amikin)
am-eh-KAY-sin (AM-eh-kin)

Some internet sources say that a "cin, c-i-n" ending means an aminoglycoside, but that is not necessarily true. Many antibiotics end in "c-i-n," so I think it's more useful to look at the "a-m-i" that is in **aminoglycoside** (*the drug class*), **amikacin** (*the generic name*) and **Amikin (the brand name)**.

The brand name **Amikin** is simply **amikacin** without the "a-c." **Amikin**.

2) Genta**micin** (brand, Garamycin)
Jenn-ta-MY-sin (gare-uh-MY-sin)

Just as practitioners abbreviate **vancomycin** as "vanc" in conversation, they abbreviate **gentamicin** as "gent." The brand name **Garamycin** is similar to **gentamicin** spelled with "mycin, m-y-c-i-n" not "micin, m-i-c-i-n". This World Health Organization frowns on brand names that have confusing stem-like parts to them.

IMMUNE SECTION V. Antibiotics for urinary tract infections (U-T-Is) and peptic ulcer disease (P-U-D)

Dihydrofolate reductase inhibitor antibiotics

Sulfamethoxazole with trimethoprim is a combination therapy that affects the folic acid in bacteria. Humans can safely ingest folic acid, so it doesn't affect us adversely.

Sulfa drugs clear urinary tract infections (UTIs) and provide prophylaxis (prevention) of certain infections that commonly occur in immunocompromised patients such as HIV patients.

However, sulfa medications can sometimes cause allergic reactions. **Sulfamethoxazole** can even cause a rare but life-threatening condition of the skin and mucous membranes known as Stevens-Johnson syndrome.

Fluoroquinolone antibiotics

We sometimes call **fluoroquinolones** "floxacins" after their infix "-fl-" + suffix "oxacin." Like tetracyclines, fluoroquinolones cause photosensitivity (an increased sensitivity to burning from sunlight) and chelation, a binding with cations such as the Ca++ in milk or antacids. **Fluoroquinolones** have a very unusual side effect in that sometimes they can cause tendon rupture, although rarely.

Nitroimidazole antiprotozoal

Metronidazole treats various infections, including *H. Pylori*, as part of triple therapy. A notable side effect of **metronidazole** is the disulfiram reaction where a patient may experience serious nausea and vomiting. Projectile vomiting is rare, but it is a vivid way to remember **metronidazole's** adverse effect with alcohol consumption.

Let's take a closer look at the drugs names and stems.

DIHYDROFOLATE REDUCTASE INHIBITOR COMBINATION ANTIBIOTIC

Sulfamethoxazole with Trimethoprim (brand, Bactrim)
sull-fa-meth-OX-uh-zol e /try-METH-oh-prim (BACK-trim)

S-M-Z forward slash T-M-P is the acronym for sulfamethoxazole with trimethoprim.

The "sulfa, s-u-l-f-a" in sulfamethoxazole and the "prim, p-r-i-m," in trimethoprim put them in the dihydrofolate reductase inhibitor class. (Remember this enzyme helps the bacteria make folic acid)

Memorizing Pharmacology

While sulfa drugs have "s-u-l-f-a" in them, some drugs have sulfa groups in the chemicals' structure, but not in the generic name, e.g. , **furosemide.**

I want to caution you about seeing sulfa moieties in chemical structures and assuming it will cause an allergic reaction.

The academic literature doesn't support cross-sensitivity between allergies to sulfa antibiotics and other sulfonamide containing drugs like **furosemide**.

The brand name contains "b-a-c-t-r-i-m" from "**bact**e**ri**u**m**" to make **Bactrim**.

LET'S LOOK AT TWO FLUOROQUINOLONE ANTIBIOTICS

1) Cipro<u>floxacin</u> (brand, Cipro)
sip-row-FLOCKS-uh-sin (SIP-row)

The **quinolone** stem is "oxacin, o-x-a-c-i-n" but **fluoroquinolones** have the "f-l" infix also.

One student remembered quinolones are for UTIs because Dr. *Quinn*, Medicine Woman, a 90's television doctor, is female. Women get proportionally more UTIs than men do.

By cutting the "floxacin" stem from the generic name **ciprofloxacin**, the manufacturer made the brand name, **Cipro**.

2) Levo<u>floxacin</u> (brand, Levaquin)
Lee-vo-FLOCKS-uh-sin (LEV-uh-Quinn)

Levofloxacin is the left-handed (levo) isomer of **ofloxacin**, another **fluoroquinolone**. The brand name combines the "lev" from <u>lev</u>ofloxacin and "quin" from fluoro**quin**olone to form **Levaquin**.

NITROIMIDAZOLE ANTIPROTOZOAL

Metro<u>nidazole</u> (brand, Flagyl)
met-ruh-NYE-duh-zole (FLADGE-ill)

The generic name **metronidazole** differs by only three letters from the word "nitroimidazole," its parent class. It's stem, "nidazole, n-i-d-a-z-o-l-e," is sometimes confused as "a-z-o-l-e" which is simply a type of chemical compound. Make sure to use the whole stem, otherwise you'll confuse the other stems that end in "a-z-o-l-e" like prazole, conazole, and piprazole.

Gastroenterologists use **metronidazole** for peptic ulcer disease (P-U-D). Note that **metronidazole** is technically an antiprotozoal.

Some students look at the "azole" (a-z-o-l-e) ending not as a stem, but for its similarity to "ozoal" (o-z-o-a-l) from "protozoal."

Another student learned to give "Flag a shorter form of metronidazole's brand **Flagyl**, for *B. frag*, a shortening of the *Bacteroides fragilis* infections.

IMMUNE SECTION VI. ANTI-TUBERCULOSIS AGENTS

Prescribers use anti-tuberculosis agents for an extended period (several months) because tuberculosis organisms grow slowly. Multiple drug therapy helps prevent resistance. I use the acronym, ripe, "r-i-p-e," to remember the four major antituberculosis agents: **<u>r</u>ifampin (R)**, **<u>i</u>soniazid (I)**, **<u>p</u>yrazinamide (P)**, and **<u>e</u>thambutol (E)**.

Non-drug resistant, non-HIV patients take all four drugs for two months, and then **isoniazid** and **rifampin** together for four more months.

Let's look a little more closely at the four antituberculars.

1) Rifampin (brand, Rifadin)
rif-AM-pin (rif-UH-din)

Students remember that **rifampin** turns secretions like tears, sweat, and urine red with its first letter "r." The brand name **Rifadin** simply replaces the "m-p" from **rifampin** with a "d" to get Rifadin.

2) Isoniazid (brand, INH)
eye-sew-NIGH-uh-zid (EYE-en-aitch)

The "n-i" in the middle of **isoniazid** reminds students that peripheral **n**eur**i**tis is an adverse effect, taking the "n-i" from neuritis.

There is no "H" in **isoniazid**, so the brand name **I-N-H** comes from those letters in the chemical name **i**so**n**icotinyl-**h**ydrazide.

3) Pyrazinamide (brand, P-Z-A)
pier-uh-ZIN-uh-mide (pee-zee-ay)

The "p" in **pyrazinamide** reminds students that an adverse effect is polyarthritis. Pulling letters from within the generic name, **py**ra**z**in**a**mide's abbreviation is **P-Z-A**.

4) Ethambutol (brand, Myambutol)
eh-THAM-byou-tall (my-AM-byou-tall)

The "e" for "eyes" or "o" in **ethambutol** helps remind students of vision and optics, and that optic neuritis is an adverse effect.

To make the brand name, the manufacturer replaced the "eth" in **ethambutol** with "my." The m-y begins **Myambutol** because *Mycobacterium tuberculosis* is the causative agent.

IMMUNE SECTION VII. ANTIFUNGALS

Scientists divide antifungals into two general types: systemic (in the body) and dermatologic or topical (on the skin). Before the advent of antifungals, most systemic fungal infections were deadly.

Amphotericin B can treat systemic infections, sometimes referred to as "ampho-terrible" because of its side effects.

Fluconazole orally treats vaginal yeast infections.

Nystatin can eliminate thrush or yeast infections.

Let's look at these three antifungal names more closely.

1) Amphotericin B (brand, Fungizone)
am-foe-TER-uh-sin bee (FUN-gah-zone)

What about **amphotericin A**? Well, it didn't do anything, so they came up with **amphotericin B**. While the antibacterials' brand names didn't do a very good job helping to indicate their therapeutic effects, this antifungal's brand name, **Fungizone**, makes it easier to know its therapeutic use.

2) Flu<u>conazole</u> (brand, Diflucan)
flue-CON-uh-zole (die-FLUE-can)

The "conazole, c-o-n-a-z-o-l-e" ending helps identify **flu<u>conazole</u>** as an antifungal drug. Again, be careful of the "f-l-u" in **fluconazole**, which is for a <u>flu</u>orine atom it contains, not influenza.

One student came up with using the first three letters of the brand name **Diflucan** as the vehement command, "<u>Di</u>e <u>f</u>ungi!"

3) Nystatin (brand, Mycostatin)
NIGH-stat-in (MY-co-stat-in)

Nystatin is an interesting generic name because it ends in "statin." A class of cholesterol lowering drugs, the HMG-CoA reductase inhibitors, commonly referred to as "statins," have a similar ending.

A better infix + suffix stem for HMG-CoA reductase inhibitors is "vastatin, v-a-s-t-a-t-i-n."

To keep from thinking **nystatin** was ever a cholesterol lowering "statin," one student remembered the dosage forms nystatin comes in: powder and liquid to swish and spit.

Mycostatin, the brand name, dropped the "ny, n-y" from **ny̲statin** and added "Myco, m-y-c-o" a prefix often seen with my̲co̲ses (fungal infections).

IMMUNE SECTION VIII. ANTIVIRALS - NON-HIV

Many antivirals have "-vir-" (v-i-r) in the middle or at the end of the generic and or brand name.

Drugs for influenza, such as **oseltamivir** and **zanamivir** work when taken within 48 hours of the infection.

Drugs for herpes infections such as **acyclovir** and **valacyclovir** can help prevent recurrences and treat an infection, but they do not cure the disease.

Respiratory syncytial virus (RSV) is usually unproblematic in healthy adults, but in infants younger than one year old, it can be deadly. A drug like the vaccine **palivizumab** can prevent RSV in at-risk patient populations.

TWO ORAL INFLUENZA A AND B ANTIVIRALS

1) Oselt̲amivir (brand, Tamiflu)
owe-sell-TAM-eh-veer (TA-mi-flue)

Recognize oseltamivir's drug class from the amivir, a-m-i-v-i-r stem. Often family members will all get prescriptions for **oseltamivir** if one person is sick enough or if a family member is immunocompromised. The brand name **Tamiflu** alludes to a drug that "tames the flu." It's prescribed for acute influenza or prophylaxis.

2) Zan<u>amivir</u> (brand, Relenza)
za-NAH-mi-veer (rah-LEN-zuh)

Also recognize zanamivir's drug class from the amivir, a-m-i-v-i-r stem. **Zanamivir** comes in a Diskhaler, a way to get powder to the lungs. The Diskhaler is difficult for patients with dexterity issues, but provides an alternative to **oseltamivir (brand, Tamiflu).**

Think: **Relenza** "<u>re</u>presses influ<u>enza</u>" virus, or **Relenza** makes "influ<u>enza</u> <u>rel</u>ent" (give up).

Two Herpes simplex virus (HSV) & Varicella-Zoster virus (VZV) antivirals

1) A<u>cyclovir</u> (brand, Zovirax)
ay-SIGH-clo-veer (zo-VIE-racks)

Zovirax treats Varicella-Zoster virus (VZV) and herpes simplex virus (HSV).

You can think of **Zovirax** as a drug that <u>ax</u>es <u>Z</u>oster <u>vir</u>us.

Dosing is five times daily.

2) Vala<u>cyclovir</u> (brand, Valtrex)
Val-uh-SIGH-clo-veer (VAL-trex)

Val<u>acyclovir</u> has **acyclovir** in the name because it's the valine ester. A prodrug like **valacyclovir** turns into an active drug in the body, in this case, **acyclovir**.

Valacyclovir allows for twice daily dosing, so prescribers prefer the oral form to five times daily dosing of **acyclovir** for patient compliance.

The brand name, **Valtrex,** includes the "val" from **<u>val</u>acyclovir** plus T-rex, and wrecks a virus; Val-T-Rex... Valtrex.

Memorizing Pharmacology

RESPIRATORY SYNCYTIAL VIRUS (RSV)

Pali<u>vi</u>zumab (brand, Synagis)
pal-eh-viz-YOU-mab (SIN-uh-giss)

The prefix "p-a-l-i" has "p" and "i" in it. You can remember **palivizumab** is for <u>p</u>ediatrics or <u>i</u>nfants at risk for RSV.

In **palivizumab**, the "pali" is a prefix that separates it from similar drugs.

The "-vi-" (v-i) stands for antiviral (the target),
the "-zu-" (z-u) stands for humanized (the source),
and the "-mab" (m-a-b) is for <u>m</u>onoclonal <u>antib</u>ody.

This biologic stem + infixes resemble **infliximab (brand, Remicade)** for ulcerative colitis or **omalizumab (brand, Xolair)** for asthma, but with a different clinical purpose.

IMMUNE SECTION IX. ANTIVIRALS - HIV

HIV drugs affect specific targets in the cell or retrovirus. HIV medications, like tuberculosis medications, work best in drug combinations.

I've organized five HIV drug classes in the order an HIV virus attacks a healthy cell.

First, the HIV virus tries to *fuse* with the cell, then it uses cellular chemokine receptor five (CCR5) to enter the cell. Inside the cell, the HIV virus uses reverse transcriptase, integrase, and then protease.

HIV medications have three letter abbreviations, as these drugs are not only hard to pronounce, but conversation filled with several multisyllable words can make comprehension difficult.

Because these generic drug names are so hard to memorize, I have placed the brand name mnemonic first, then the details on the generic stems.

1) Fusion inhibitor

Enfu<u>vir</u>tide (brand, Fuzeon) (ENF, T-20)
En-FYOO-veer-tide (FYOO-zee-on)

It's easier to remember **enfuvirtide's** brand name **Fuzeon** first because it's a <u>fu</u>si<u>on</u> inhibitor, taking the f-u and o-n from fusion; **Fuzeon**.

Inside the generic name, you see "vir" (v-i-r) for antiviral and we pronounce the "f-u" as FYOO.

Put that together and you can remember **enfuvirtide** is brand name **Fuzeon**, a fusion inhibitor.

I use the "T" in "**T-20**" to remember that "tide" is the last syllable in the generic name.

2) Cellular chemokine receptor (CCR5) antagonist

Mara<u>viroc</u> (brand, Selzentry) (MVC)
MARE-uh-VIR-ock (SELLS-en-tree)

The brand name **Selzentry** sounds a lot like "sentry," someone who guards.
The stem "vir, v-i-r" is inside the generic name **mara<u>vir</u>oc**. The sub-stem "viroc, v-i-r-o-c" has five letters, with the "c" at the end of the generic name, so you can remember the two "c's" in CCR5 antagonist.
You can think of **maraviroc** as a "<u>roc</u>k" guarding against viral entry.

3) One Non-nucleoside reverse transcriptase inhibitor (N-N-R-T-I) with 2 Nucleoside forward slash Nucleotide reverse transcriptase inhibitors (N-R-T-Is)

Efa<u>vir</u>enz with Emtri<u>cit</u>abine and Tenofo<u>vir</u> (brand, Atripla) (EFV with FTC and TDF)

65

eh-FAH-vir-enz / EM-try-SIGH-tah-been / ten-OFF-oh-vir (ay-TRIP-lah)

The brand name **Atripla** can be thought of as three drugs, "triple" surrounded by two A's that can stand for "against AIDS."

When you see something complex to memorize, first look at the sub-class of antiviral.

Instead of trying to memorize the whole name, try to memorize the stems "virenz, v-i-r-e-n-z," "citabine, c-i-t-a-b-i-n-e," and "vir, v-i-r".

Then add the other two or three syllables to memorize the whole names of **efavirenz, emtricitabine**, and **tenofovir**.

4) INTEGRASE STRAND TRANSFER INHIBITOR

Raltegravir (brand, Isentress) (RAL)
ral-TEG-ra-veer (EYE-sen-tress)

The brand name **Isentress** also looks like sentry, except it has the "I" to remind you of the integrase enzyme.
Raltegravir is an integrase strand transfer inhibitor. Inside the generic name, you can find the stem "tegravir, t-e-g-r-a-v-i-r" made up of "tegra, t-e-g-r-a" a part of integrase and "vir, v-i-r" for antiviral.

5) PROTEASE INHIBITOR

Darunavir (brand, Prezista) (DRV)
dar-YOU-nah-veer (Pre-ZIST-uh)

The brand **Prezista** sounds like resist if it was spelled "r-e-z-i-s-t," and the first two letters "p-r" of "protease."
The "navir, n-a-v-i-r" stem indicates protease inhibitor.

Chapter 4 Quiz: Immune, Drug stem and drug classification practice

I'm going to read a generic name for you and I want you to think about the stem, if any, and the class of medication. Note: in this group of medications, the stem mycin, m-y-c-i-n, and micin, m-i-c-i-n alone don't indicate a specific class of medication, so just a reminder mycin might hint at antibacterial, but that isn't definitive.

1. Neo<u>mycin</u> with Polymyxin-B and Bacitracin, mycin, antibacterial
2. Butenafine, no stem, antifungal
3. Influenza vaccine, no stem, prophylactic antiviral
4. Docosanol, no stem, acute antiviral
5. Amoxicillin, cillin, c-i-l-l-i-n, penicillin class antibiotic
6. Amoxicillin with clavulanate, cillin, c-i-l-l-i-n, penicillin class antibiotic and beta-lactamase inhibitor
7. Cephalexin, ceph, c-e-p-h, 1st generation cephalosporin
8. Ceftriaxone, cef, c-e-f, 3rd generation cephalosporin
9. Cefepime, cef, c-e-f, 4th generation cephalosporin
10. Vancomycin, mycin, m-y-c-i-n, glycopeptide antibiotic
11. Doxycycline, cycline, c-y-c-l-i-n-e, tetracycline antibiotic
12. Minocycline, cycline, c-y-c-l-i-n-e, tetracycline antibiotic
13. Azithromycin, infix "thro" plus stem "mycin," thromycin, t-h-r-o-m-y-c-i-n, macrolide antibiotic
14. Clarithromycin, infix "thro" plus stem "mycin," thromycin, t-h-r-o-m-y-c-i-n, macrolide antibiotic
15. Erythromycin, infix "thro" plus stem "mycin," thromycin, t-h-r-o-m-y-c-i-n, macrolide antibiotic

Memorizing Pharmacology

16. Clindamycin, mycin, m-y-c-i-n, lincosamide antibiotic
17. Linezolid, zolid, z-o-l-i-d, oxazolidinone antibiotic
18. Amikacin, kacin, k-a-c-i-n, aminoglycoside antibiotic
19. Gentamicin, micin, m-i-c-i-n, aminoglycoside antibiotic
20. Sulfamethoxazole with Trimethoprim, sulfa, s-u-l-f-a and prim, p-r-i-m, dihydrofolate reductase inhibitors
21. Ciprofloxacin, infix f-l plus stem oxacin, floxacin, f-l-o-x-a-c-i-n, fluoroquinolone antibiotic
22. Levofloxacin, infix f-l plus stem oxacin, floxacin, f-l-o-x-a-c-i-n, fluoroquinolone antibiotic
23. Metronidazole, nidazole, n-i-d-a-z-o-l-e, nitroimidazole antiprotozoal also for peptic ulcer disease
24. Rifampin, rifa, r-i-f-a, anti-tuberculosis, "R" in antituberculosis mnemonic "RIPE"
25. Isoniazid, no stem, anti-tuberculosis, "I" in antituberculosis mnemonic "RIPE"
26. Pyrazinamide, no stem, anti-tuberculosis, "P" in antituberculosis mnemonic "RIPE"
27. Ethambutol, no stem, anti-tuberculosis, "E" in antituberculosis mnemonic "RIPE"
28. Amphotericin B, no stem, antifungal
29. Fluconazole, conazole, c-o-n-a-z-o-l-e, antifungal
30. Nystatin, no stem, antifungal
31. Oseltamivir, amivir, a-m-i-v-i-r, antiviral for influenza
32. Zanamivir, amivir, a-m-i-v-i-r, antiviral for influenza
33. Acyclovir, cyclovir, c-y-c-l-o-v-i-r, antiviral for herpes simplex virus (H-S-V) and varicella-zoster virus (V-Z-V)
34. Valacyclovir, cyclovir, c-y-c-l-o-v-i-r, antiviral for herpes simplex virus (H-S-V) and varicella-zoster virus (V-Z-V)
35. Palivizumab, vizumab, v-i-z-u-m-a-b, for respiratory syncytial virus (R-S-V)

CHAPTER 4: IMMUNE

36. Enfuvirtide, vir, v-i-r, HIV antiviral fusion inhibitor
37. Maraviroc, viroc, v-i-r-o-c, HIV antiviral cellular chemokine receptor (C-C-R-5) antagonist
38. Efavirenz with Emtricitabine and Tenofovir, virenz, v-i-r-e-n-z, HIV antiviral non-nucleoside reverse transcriptase inhibitor, N-N-R-T-I, with citabine, c-i-t-a-b-i-n-e, and vir, v-i-r, HIV antiviral nucleoside nucleotide reverse transcriptase inhibitors (N-R-T-eyes)
39. Raltegravir, tegravir, t-e-g-r-a-v-i-r, HIV antiviral integrase strand transfer inhibitor
40. Darunavir, navir, n-a-v-i-r, HIV antiviral protease inhibitor

IMMUNE: MEMORIZING THE CHAPTER

"I" IMMUNE

The sub-algorithm is to start with antibacterial agents, then go to antifungals, then antivirals, which are in alphabetical order

b for bacterial
f for fungal
and v for viral
of what they are "anti-;" anti-bacterials, anti-fungals, anti-virals.

So first the OTC antibacterial cream **neomycin with polymyxin-B and bacitracin; butenafine**, the antifungal cream; and the prophylactic antiviral **influenza vaccine**; followed by an acute antiviral – **docosanol**.

Begin again with systemic antibacterials, first those that attack the cell wall, the beta lactamase susceptible amino penicillin **amoxicillin**, followed by the augmented beta lactamase resistant **amoxicillin** with **clavulanate**.

69

Memorizing Pharmacology

Move to cephalosporins in generational order: 1st generation **cephalexin**; to 3rd generation **ceftriaxone**;

to 4th generation **cefepime**, the "maximum" generation; to alphabetically last "v" for the glycopeptide antibiotic **vancomycin** for MRSA.

From bactericidal cell wall inhibitors to bacteriostatic inhibitors of protein synthesis, the tetracyclines: **doxycycline** and **minocycline** followed by three macrolides in order of the number of times taken per day: qd (once daily), b-i-d, (twice daily), q-i-d, (four times daily), **azithromycin, clarithromycin**, and **erythromycin** respectively.

Use the "-mycin" ending to get to **clindamycin** and the "l-i-n" from *clin*damycin to go to *lin*e**zolid**.

Follow with the bactericidal inhibitors of protein synthesis, the aminoglycosides **amikacin** and **gentamicin**.

Move to the UTI medications **sulfamethoxazole** and **trimethoprim, ciprofloxacin**, and **levofloxacin;** and from "l" in "levo" to "m" in **metronidazole**, the antiprotozoal.

Then four letters for four TB drugs because over 4 mm induration is a positive TB test or a ripe, "r-i-p-e" result: (R) in **rifampin**, (I) in **isoniazid**, (P) in **pyrazinamide**, and (E) in **ethambutol**.

TB patients often have opportunistic fungi, so three antifungals alphabetically follow: **amphotericin B, fluconazole,** and **nystatin.**

The antivirals for influenza follow those: **oseltamivir** and **zanamivir**; then **acyclovir** and **valacyclovir** for HSV and VZV, in order of half-life; then for RSV **palivizumab**; then for HIV, in order of attack: fusion, CCR5, reverse transcriptase, integrase, and protease. I have to use brand names to memorize these first because those are easier: **Fuzeon, Selzentry, Atripla, Isentress, Prezista**, and then I attach the generic names for those: **enfuvirtide, maraviroc,** (**efavirenz** with **emtri<u>ci</u>tabine** and **tenofo<u>vir</u>), ralte<u>gravir</u>** and **da-ru<u>nav</u>ir**.

Can you recite these in order?

Neomycin with
  Polymyxin-B and
  Bacitracin
	Azi<u>thromycin</u>
	Clari<u>thromycin</u>
	Ery<u>thromycin</u>
		Amphotericin B
		Flu<u>cona</u>zole
		Nystatin

Butenafine	Clinda<u>mycin</u> Line<u>zolid</u>	Osel<u>tami</u>vir Zanamivir
Influenza vaccine Docosanol	Ami<u>kac</u>in Genta<u>micin</u>	A<u>cyc</u>lovir Vala<u>cyc</u>lovir
Amoxi<u>cillin</u> Amoxi<u>cillin</u> with Clavulanate	Sul<u>fa</u>methoxazole with Trimetho<u>prim</u>	Pali<u>vi</u>zumab Enfu<u>vir</u>tide Mara<u>vir</u>oc

Ce<u>ph</u>alexin
Ce<u>ft</u>riaxone
Ce<u>f</u>epime
	Cipro<u>flox</u>acin
	Levo<u>flox</u>acin
	
	Metro<u>nida</u>zole
		Efa<u>vir</u>enz with
		  Emtri<u>ci</u>tabine
		  and
		  Tenofo<u>vir</u>
		Raltegravir

Vanco<u>mycin</u> Doxy<u>cycline</u> Mino<u>cycline</u>	Ri<u>f</u>ampin Isoniazid Pyrazinamide Ethambutol	Daru<u>nav</u>ir

Once you're able to recite the list, you are ready to move on.

CHAPTER 5: NEURO

NEURO SECTION I. OTC LOCAL ANESTHETICS AND ANTIVERTIGO

Before I get into the drugs you're most familiar with or would attach to the larger neuro class, for example, the antidepressants and antianxiety medicines, I want to talk about a few over-the-counter products that fit into this category starting with two local anesthetics and one antivertigo medication.

There are two major classes of local anesthetics named after the molecules in the middle of their structures: esters and amides. What makes these drugs special, is that they stop axonal conduction. If you go back to your anatomy class, remember the axons are the long attachments to tree-like dendrites. Most neuro drugs work on a neurotransmitter like serotonin or norepinephrine. In contrast, local anesthetics work directly on axonal conduction.

Manufacturers put *ester* type anesthetics, such as **benzocaine,** in *topical* agents because when we give ester type anesthetics by injection, they are more allergenic or they cause allergic reactions.

Amides are less allergenic; therefore, patients are less likely to have an issue when we inject the *amide*-type local anesthetic **lidocaine**.

Most students know of **cocaine** and remember these amides are local anesthetics through the "–caine" ending association among the three names: **benzocaine**, **lidocaine**, and cocaine.

The over-the-counter (OTC) *antiemetic* and motion sickness medicine **meclizine,** brand **Dramamine,** also has a brand name **Antivert**, for anti-vertigo, which helps with memorization of this product.

Let's look at little more closely at these three over the counter drug names.

LOCAL ANESTHETIC - ESTER TYPE

Benzocaine (brand, Anbesol)
BEN-zoh-cane (ANN-buh-sawl)

The "caine, c-a-i-n-e" stem indicates **benzocaine** is a local anesthetic.

Anbesol has the letter "n" and the letter "b" in it, the first and last letter of the word numbs for its use in numbing an aching tooth.

LOCAL ANESTHETIC - AMIDE TYPE

Lidocaine (OTC brand, Solarcaine)
LIE-doe-cane (SOH-ler-cane)

We often use **lidocaine** topically over-the-counter to treat sunburns, hence the brand name **Solarcaine.**

Injectable and patch forms of **lidocaine** are also available by prescription. Paramedics often use **lidocaine** for arrhythmias in emergencies as an injectable, which is part of the Lean, L-E-A-N acronym for the emergency medicines: **lidocaine**, **epinephrine**, **atropine**, and **naloxone**.

ANTIVERTIGO

Meclizine (brand, Dramamine)
MECK-luh-zeen (DRAH-mah-mean)

If you squish the "c" and "l" in meclizine to make a "d" and use the ensuing "i-z-i," you get d-i-z-i, dizzy. **Antivert** is an alternate brand name to Dramamine and forms part of the word **antivert**igo.

There's one more over-the-counter combination medication I want to talk about before we segue into the prescription-only products and that's acetaminophen with diphenhydramine, better known stateside as brand Tylenol P-M.

NEURO SECTION II. SEDATIVE-HYPNOTICS (SLEEPING PILLS)

A patient came into the pharmacy saying his wife wanted **Tylenol P-M**, but saw how expensive the brand name product was.

I told him that **Tylenol P-M** contains **acetaminophen** and **diphenhydramine** and that he could buy both generics separately and it would be cheaper in the end.

He thought about it a minute and said, "It might be a little cheaper at first, but when my wife sends me back to get **Tylenol P-M**, what she asked for, it might not be cheaper after all."

This story highlights the importance of reading O-T-C labels closely. It's an opportunity for practitioners to help patients understand OTC products whether they take the advice or not.

The academic literature supports that patients tend to buy more brand name products *without* the pharmacist's help and more generic products *with* the pharmacist's help. I think there are at least four reasons for this:

First, I think patients don't trust their own understanding of the multi-syllable generic names.

Second, brand name boxes and products are much brighter and more colorful than their generic counterparts are.

You can think of seeing two Mallard ducks on a pond, a male and a female. The drake, the male, has a yellow bill, bright green head, white neckline, brown chest, and kind of a grayish body.

The female feathers display different shades of drab brown. My understanding is that this is protective so the female blends in better and is more likely to survive an attack.

However, the generic box, with minimalist coloring and lettering, can provide the exact same effect as the brand at a lower price, but it's a bit camouflaged and a person might need a guide, like a pharmacist, to help them find it and or trust it.

Third, we read English from left to right, so we'll see the brand name product first as it's place on the pharmacy shelf, by convention on the left hand side. It's just like being in a hurry during an exam, you hit the right answer, and you bubble in the letter and move on ignoring the other possibilities.

Fourth, brand names are easier to say. It's a lot easier to ask someone to get a bottle of Tylenol P-M, than it is to say, get some acetaminophen with diphenhydramine. Although the person might be perfectly happy with the generic, they didn't ask for it directly.

Let's talk a little bit about sedative-hypnotic medications and how their names correspond to the indication.

Manufacturers often represent the O-T-C sedative **diphenhydramine** with the P-M acronym (just a reminder, P-M stands for post-meridiem (the Latin) for after mid-day) but in practice, it really stands for evening or nighttime.

Prescription sedative-hypnotics also provide hints about their function in their brand names:

Eszopiclone (brand, <u>Lunesta</u>) contains Luna for "moon,"

Ramelteon, brand name Roze*rem* refers to rapid eye movement (R-E-M, REM) sleep,

and **zolpidem's brand name Ambien** creates an "ambient" or tranquil environment.

Although the oral benzodiazepines **alprazolam, clonazepam, diazepam,** and **lorazepam** can all work as sedative hypnotics, I discuss benzodiazepines, benzos, later in this chapter as they have other functions as well.

OTC - Non-narcotic analgesic with Sedative-Hypnotic

Acetaminophen with Diphenhydramine (brand, Tylenol P-M)
uh-seat-uh-MIN-no-fin / dye-fen-HIGH-dra-mean
(TIE-len-all pee em)

It's common to combine two drugs like **acetaminophen** for aches and **diphenhydramine**, a sedating 1st- generation antihistamine. One of my students came up with mnemonic to take both "phens," acetamino*phen* and di*phen*hydramine when you want to end up sleepin'.

Two Benzodiazepine-like sedative hypnotics

1) Eszopi*clone* (brand, Lunesta)
es-zo-PEH-clone (Lou-NES-tuh)

Eszopiclone's generic stem "-clone, c-l-o-n-e" will put you in the sleeping zone is how one student remembered eszopiclone's function.

Some students point to the "z" in es**z**opiclone for getting your z's.

The brand name **Lunesta** uses the Latin for moon (Luna) plus part of the word "rest," which combines a visual with therapeutic function. **Lunesta**.

2) Zolpidem (brand names, Ambien or Ambien CR)
zole-PEH-dem (AM-bee-en)

Use the "-pidem, p-i-d-e-m" stem to remember **zolpidem** as a sedative-hypnotic. Some students point to the "z" in **zolpidem** for getting z's just like eszopiclone.

Some students match the brand name **Ambien** with an ambient, sleepy environment.

Zolpidem Controlled Release, brand **Ambien C-R**, works for people who have difficulty maintaining sleep (D-M-S) *and* those who have difficulty falling asleep (D-F-A).

The regular zolpidem only works for those with difficulty falling asleep, D-F-A.

A MELATONIN RECEPTOR AGONIST SEDATIVE HYPNOTIC

Ramelteon (brand, Rozerem)
ra-MEL-tee-on (row-ZER-em)

The "–melteon, m-e-l-t-e-o-n" stem in **ramelteon** lets you know it's a melatonin agonist.

A student said the "m-e-l" in **ramelteon** reminded her of the word mellow, m-e-l-l-o-w

In **Rozerem**, you can see the "z" for z's, the "r-e-m" for rapid eye movement REM sleep and "roze" rhymes with doze.

CHAPTER 5: NEURO

NEURO SECTION III. Antidepressants

The four antidepressant classes often intimidate students, so let's take them one acronym at a time.

1) SSRIs

The <u>s</u>elective <u>s</u>erotonin <u>r</u>euptake <u>i</u>nhibitor (SSRI) class includes drugs such as **citalopram, escitalopram, sertraline, paroxetine,** and **fluoxetine.**

Note: **escitalopram** has the same "es" prefix (added onto **citalopram**) discussed with the proton pump inhibitors **esomeprazole** and **omeprazole**. Again, we expect the sinister "S" form to perform better.

These medications will *selectively inhibit reuptake* (the breakdown) of *serotonin* within the neurons. To put it in plain English, think of reuptake as recycle.

Every other week, a large truck comes by my house and picks up our recycling from a large green container. If I don't get the container to the curb, then I'm stuck with a 96-gallon cart worth of paper, cardboard boxes and aluminum cans.

SSRIs keep the serotonin from making it to the curb to be picked up and recycled. That increased serotonin level improves mood.

2) SNRIs

Similar to the S-S-R-Is are the <u>s</u>erotonin-<u>n</u>orepinephrine <u>r</u>euptake <u>i</u>nhibitors (SNRIs) **duloxetine** and **venlafaxine.** Be careful - **duloxetine** is an S-N-R-I, yet has the "oxetine" stem of some S-S-R-Is like **flu<u>oxetine</u>** and **par<u>oxetine</u>**.

3) TCAs

SSRIs and SNRIs carry the names of the neurotransmitters they affect. The tricyclic antidepressant (T-C-A) class name comes from the chemical structure's three rings. **Amitriptyline** is an example.

4) MAOIs

Another group of antidepressants includes the monoamine oxidase inhibitors (M-A-O-Is). The word "monoamine" hides the neurotransmitter names that this drug affects such as the monoamines serotonin and norepinephrine.

A word that ends in "ase, a-s-e" is usually an enzyme, so if an antidepressant blocks the enzyme monoamine *oxidase* that breaks a neurotransmitter down, then there is more neurotransmitter (the monoamines, in this case) available to elevate the patient's mood. An example of an M-A-O-I is **isocarboxazid**.

Let's go through those drug classes again, but taking a careful look at the individual drug names starting with five SSRIs.

SELECTIVE SEROTONIN REUPTAKE INHIBITORS (SSRIs)

SSRI One) Citalopram (brand, Celexa)
si-TAL-oh-pram (sell-EX-uh)

Most students, when seeing two drugs with the same root, **citalopram** and **escitalopram**, quickly put them into long-term memory.

One student's trick was to remember that the "pram, p-r-a-m" medications are for de-p-r-ession, but "pram" is not an official stem.

I associate the brand name **Celexa** with the word "relax," as some of the SSRIs are used not only for depression, but for anxiety.

SSRI two) Escitalopram (brand, Lexapro)
es-si-TAL-oh-pram (LECKS-uh-pro)

While we don't have a stem for escitalopram, if you pair this medication with citalopram and these other three SSRIs, hopefully you'll remember its function.

The brand name **Lexapro** takes the last four letters of Ce**lexa**, l-e-x-a and adds "pro, p-r-o" You can think of this as the *pro*-fessional upgrade, as **Lexapro** came after **Celexa**.

It's common for an S isomer to come to market after the original drug has become available as a generic.

SSRI three) Ser**traline** (brand, Zoloft)
SIR-tra-lean" (ZO-loft)

One should use the "-traline," t-r-a-l-i-n-e stem to remember **ser*traline*** is an SSRI.

Most students also memorize the brand name **Zoloft** first as "<u>loft</u>ing" a depressed patient's mood.

SSRI four) Flu**oxetine** (brand names Prozac or Sarafem)
flue-OX-uh-teen (PRO-Zack)

Fluoxetine was the first SSRI to make it to market. The "-oxetine," o-x-e-t-i-n-e ending is supposed to be for **fluoxetine**-like entities, but you will see "-oxetine" on the S-N-R-I antidepressant **dul*oxetine*** (brand, **Cymbalta**) and ADHD medication **atom*oxetine*** (brand, **Strattera**), so be careful with this particular stem.

When **flu*oxetine*** gained a new indication, for premenstrual dysphoric disorder (P-M-D-D), it also gained a new brand name: **Sarafem, pronounced and spelled** – "Sara,"s-a-r-a like the girl's name and "fem," f-e-m for feminine.

The highest ranked winged angels are Sera-p-h-i-m, so combatting P-M-D-D might be thought of as the work of angels. I don't know if that's what the drug manufacturer was going for, but that's how I've remembered it.

In addition, by taking a new brand name for another indication, it might have prevented the potential confusion of a patient diagnosed with depression on brand name **Prozac** and a patient with P-M-D-D on brand name **Sarafem**.

The combination of "pro, p-r-o" for positive and the strong sounding "zac, z-a-c" ending makes **Prozac** sound like a strong antidepressant.

SSRI five) Par<u>oxetine</u> (brand names Paxil or Paxil C-R)
par-OX-eh-teen (PACKS-ill)

Paroxetine is similar to the SSRI **fluoxetine** with the same "oxetine" stem. **Paxil** takes "p-a-x"' from **pa**roxetine.

The controlled-release C-R version of **Paxil** is supposed to have fewer initial side effects and be a little easier to dose.

Let's look at a related class of antidepressant by examining two S-N-R-I antidepressants.

Two Serotonin-norepinephrine Reuptake Inhibitors (SNRIs)

SNRI one) Dul<u>oxetine</u> (brand, Cymbalta)
doo-LOX-eh-teen (SIM-bal-tah)

While **duloxetine** ends in "oxetine," it is an SNRI, not SSRI. **Duloxetine** affects serotonin *and* norepinephrine and one can think of the "du" as duo for two. It's also pronounced like the French deux, d-e-u-x which also means two.

I have never seen an unhappy cymbal player in a band and "alta" means tall in Spanish. So, I picture a happy, tall cymbal player to remember brand **Cymbalta** elevates mood.

SNRI two) Venla<u>faxine</u> (brand, Effexor)
ven-luh-FAX-een (Eff-ECKS-or)

We can best memorize this drug by its stem "-faxine, f-a-x-i-n-e" as there is a related medication, **Desvenla<u>faxine</u>** (brand, **Pristiq**) out there. If you look at the "afax, a-f-a-x" in venl<u>afax</u>ine and "Effex, E-f-f-e-x" in brand name **Effex**or, you can see some commonalities as to this drug having antidepressant effects.

We've just covered 7 antidepressants named after their neurotransmitters, let's look at a drug named after its shape and another after the enzyme it affects.

Tricyclic antidepressant (TCA)

Ami_triptyline_ (brand, Elavil)
ah-meh-TRIP-ta-lean (ELLE-uh-vill)

Amitriptyline's stem, is "triptyline, t-r-i-p-t-y-l-i-n-e," and I remember it's spelled t-r-*i*-p, then t-*y*-l, as this med trips up depression.

Using the "tri" in **ami_tri_ptyline** helps students remember this is a "T-C-A," or *tri*cyclic antidepressant.

Think of the brand **Elavil** as elevating the patient's mood.

Monoamine oxidase inhibitor (MAOI) antidepressant

Isocarboxazid (brand, Marplan)
iso-car-BOX-uh-zid (MAR-plan)

A student came up with **Marplan** for the atypical sad man who laments, "I so carve boxes" for **isocarboxazid**.

In addition, you can take the "m" and "a" from brand **Marplan** to remember it's an M-A-O-I.

NEURO SECTION IV. SMOKING CESSATION

There are many nicotine replacement products on the market. I've chosen to focus on two oral tablet forms used by prescription that help patients quit smoking: **bupropion** and **varenicline**.

Bupropion (brand names Wellbutrin or Zyban)
byoo-PRO-pee-on (well-BYOO-trin, ZY-ban)

Bupropion was first marketed as brand **Wellbutrin**, an atypical antidepressant, one that didn't fit into the SSRIs, SNRIs, TCAs, or MAOIs.

Reports must have come in that patients stopped smoking while taking it, so the company repackaged the drug with a new brand name as **Zyban**, to put a "ban" on smoking.

There is some risk with this medication, especially in patients with a history of seizures.

Varenicline (brand, Chantix)
Vah-WREN-eh-clean (CHAN-ticks)

Varenicline has the "nicline, n-i-c-l-i-n-e" stem. And if you replace that "l" with an o-t, you get n-i-c-o-t-i-n-e, nicotine.

A student, in a southern drawl, remembered varenicline's generic name by saying, "With **varenicline**, 'I'm vary incline ta quit.'"

Another came up with "My chant is, 'I don't need my fix' with **Chantix**."

NEURO SECTION V. BENZODIAZEPINES

Benzodiazepines relieve anxiety, insomnia and muscle spasms. Like the tricyclic antidepressants, benzodiazepines get their name from their chemical structure: a benzene ring and a diazepine ring.

Because benzodiazepine has many syllables, most people call them benzos. Examples include **alprazolam, midazolam, clonazepam, diazepam** and **lorazepam.**

Note benzodiazepines have the similar generic suffixes "azolam, a-z-o-l-a-m" and "azepam, a-z-e-p-a-m"

Benzos replaced barbiturates as a sleep aid because barbiturates can cause respiratory depression leading to death.

A student remembered this danger by thinking of barbiturates, which are often abbreviated barbs, as literal barbs on razor wire fences that punctured lungs representing respiratory depression. Let's look at the names of four specific benzodiazepines.

1) Alprazolam (brand, Xanax)
Al-PRAY-zo-lam (ZAN-ax)

You should remember benzodiazepines by their two endings, "azolam, a-z-o-l-a-m" or "azepam, a-z-e-p-a-m" Be careful as a number of online and highly regarded test prep resources, including those by licensed professionals, refer to benzodiazepine stems as "pam, p-a-m" or "lam, l-a-m"

That is incorrect. This will lead you to think drugs that are not benzos are in the class. For example, **vera*pam*il (brand, Calan)** is a calcium channel blocker used for hypertension and ***lam*otrigine (brand, Lamictal)** is an antiepileptic.

I've even seen a well-regarded resource indicate that **citalopram** has a "pam" suffix at the end, and that's understandable as some patients drop the "r," pronouncing it "citalopam," but this is incorrect. Citalopram is an SSRI with anti-anxiety properties, but not a benzo.

A student came up with this rhyme to connect the generic and brand name.

Alprazolam has one z; benzodiazepine has two; **Xanax** sounds like a "z" to help you get a snooze and "x's" out a**nx**iety too.

You can also look at the word **Xanax**, **X-a-n-a-x** and see the "a-n-x" from anxiety; **Xanax**.

2) Mid**azolam** (brand, Versed)
meh-DAZE-oh-lam (VER-said)

You should remember **midazolam** through the "azolam, a-z-o-l-a-m" stem, but there are other tips you can use.

There are two m's in **midazolam** for the memories you can't form since **midazolam** causes anterograde amnesia.

Just as your *ante*brachium is your forearm, and the *ante* is the money poker players put out before the dealer deals, *ante*rograde amnesia is the inability to form future memories. Alternatively, you can use the brand name; I can't remember the "verse you just said" for brand **Versed**.

3) Clon**azepam** (brand, Klonopin)
kloe-NAZ-uh-pam (KLON-uh-pin)

Clonazepam should be remembered from the "azepam, a-z-e-p-a-m" stem.

The brand name **Klonopin** uses the phonetic spelling of generic **clonazepam's** first four letters "c-l-o-n."

4) Lor**azepam** (brand, Ativan)
lore-A-zeh-pam (AT-eh-van)

Remember **lorazepam** through the "–azepam" stem or think about the brand name **Ati-*van*** **van**quishing anxiety.

NEURO SECTION VI. ATTENTION DEFICIT HYPERACTIVITY DISORDER (A-D-H-D) MEDICATIONS

A-D-H-D stimulants calm a patient who has a hyperactive mind and or body without a sedative effect.

Examples include the drugs **dexmethylphenidate** and **methylphenidate**. These two medications have the same root, **methylphenidate**. How is this different from our es-, e-s, examples from before?

In chemistry, compounds direct plane-polarized light to either the left or right. These terms are "d" or "(plus)" for dextrorotatory compounds rotating plane-polarized light to the right, and "l" or "(minus)" for levorotatory compounds rotating plane-polarized light to the left.

Dexmethylphenidate rotates plane-polarized light to the right. **Dexmethylphenidate** *should* be more effective, last longer, and have fewer side effects than methylphenidate.

Atomoxetine is a non-stimulant medication and because there's not a potential for abuse, it doesn't carry a D-E-A schedule. It's *not* an S-S-R-I like **fluoxetine**, even though it ends in the "oxetine" stem.

Let's look more closely at these drug names.

TWO STIMULANT DEA SCHEDULE II ADHD MEDICATIONS

1) Dexmethylphenidate (brand, Focalin)
dex-meth-ill-FEN-eh-date (FOE-ca-lin)

When I was a student, I could remember that **Focalin** and **Concerta** both had **methylphenidate** in their names, but I could never remember which was **dexmethylphenidate** and which was **methylphenidate**.

Memorizing Pharmacology

Then I thought of the "F" in "**Focalin**" as following the "d-e" alphabetically in **dexmethylphenidate** to help me.

To remember **Focalin** is for ADHD, you think of **Focalin** helping a patient focus.

2) Methylphenidate (brand names Ritalin and Concerta)
meth-ill-FEN-eh-date (con-CERT-uh)

Methylphenidate has many brand names, including **Ritalin** and **Concerta**. Most students seem to already know **methylphenidate,** but remember that **Concerta** can help a patient concentrate, taking c-o-n-c-e-r and t from concentrate; **Concerta**.

To remember it's an amphetamine, one student said she thought of staying up all night at a concert with **Concerta**. **Concerta** is a long-acting medication that needs to be taken only once a day.

NON-STIMULANT - NON-SCHEDULED ADHD MEDICATION

Atomoxetine (brand, Strattera)
AY-toh-mocks-e-teen (stra-TER-uh)

Note the "oxetine" stem here is not an SSRI antidepressant, but a non-stimulant medication for ADHD.

The brand name **Strattera** uses the s-t-r from straighten and a-t-t-e from attention, so straighten a patient's attention, **Strattera**.

NEURO SECTION VII. BIPOLAR DISORDER

Mood stabilizers such as **lithium** are especially likely to cause electrolyte imbalances. If you look at the periodic table, you see that lithium (L-i) and sodium (N-a) are in the same group (the alkali metals) and both have a +1 charge as an ion.

Because of this similarity, what happens to sodium will happen to **lithium,** causing either a toxic or a subtherapeutic state if too much **lithium** is retained or excreted, respectively.

Other meds, like **risperidone**, an antipsychotic, can help control certain symptoms of the disease until the **lithium** level is correct.

SIMPLE SALT - MOOD STABILIZER

Lithium (brand, Lithobid)
LITH-e-um (LITH-oh-bid)

Lithium sits in the same group on the far left of the Periodic Table of Elements as the Latin *Natrium* (N-a), commonly known as the chemical element sodium.

The body has trouble telling the difference between lithium and sodium, and too much or too little sodium can wreak havoc on **lithium** levels.

A way to remember this is the saying, "Where the salt goeth, the **lithium** goeth."

The "b-i-d" in the brand name **Lithobid** is the Latin *bis-in-die*, or twice-daily for the dosing schedule.

NEURO SECTION VIII. Schizophrenia

There are many antischizophrenic medications, how do we know which to use?

We break the medication classification for schizophrenia into typical (1st-generation) or atypical (2nd-generation). We further divide the typical antipsychotics **chlorpromazine** and **haloperidol** into low potency and high potency, respectively.

While these two antipsychotics have the same therapeutic effects, their side effect profiles are different. Low potency drugs like **chlorpromazine** cause more sedation, but fewer extrapyramidal symptoms (E-P-S). Extrapyramidal symptoms are movement disorders associated with antipsychotics.

High potency drugs like **haloperidol** cause more extrapyramidal symptoms (movement disorders), but less sedation. We prescribe typical antipsychotics for positive symptoms such as delusions, hallucinations and paranoia.

Atypical antipsychotics such as **quetiapine** and **risperidone** cause fewer extrapyramidal effects, but have more negative metabolic effects like weight gain, diabetes, and hyperlipidemia.

Second-generation drugs work for positive symptoms, such as delusions, as well as negative symptoms, such as poor motivation and emotional and social withdrawal.

Let's take a look at the drug names.

First generation antipsychotic (F-G-A) *low potency*

Chlorpromazine (brand, Thorazine)
Klor-PRO-mah-zeen (THOR-uh-zeen)

CHAPTER 5: NEURO

Chlorpromazine was the first antipsychotic. While it carries side effects, it represented a new treatment option for schizophrenic patients – a 1st generation.

Generational classifications are especially important in antipsychotics because of differences in both side effects and effects on positive versus negative symptoms.

Thor is a mythical god, and you can think of brand **Thorazine** as helping people who have delusions of mythical people.

FIRST GENERATION ANTIPSYCHOTIC (F-G-A) *HIGH POTENCY*

Halo**peridol** (brand, Haldol)
hal-low-PEAR-eh-doll (HAL-doll)

Use the "-peridol, p-e-r-i-d-o-l" stem to recognize this 1st-generation high potency drug.

Many students think of the "halo" in **haloperidol** as the halo that sits *high* on an angel's head to remember this is *high* potency.

To make the brand name **Haldol**, they just took the first and last three letters of the generic name **hal**operi**dol**; Haldol.

SECOND-GENERATION ANTIPSYCHOTICS (S-G-As) (ATYPICAL ANTIPSYCHOTICS)

Ris**peridone** (brand, Risperdal)
ris-PEAR-eh-done (RIS-per-doll)

Note the stem "–peridol, p-e-r-i-d-o-l" from **haloperidol** and "–peridone, p-e-r-i-d-o-n-e" from **risperidone** are similar; that can help you remember both of these are antipsychotics.

The brand name **Risperdal** and generic name **risperidone** share the opening letters "r-i-s-p-e-r, "risper. One student thought of risper and whisper, as in calming the whispering voices.

Quetiapine (brand, Seroquel)
Kweh-TIE-uh-peen (SEAR-uh-kwell)

In **quetiapine**, you should use the "-tiapine, t-i-a-p-i-n-e" stem to recognize that this is a 2nd-generation antipsychotic.

If you switch the "i" and "t" in **quetiapine**, q-u-e-t-i-a-p-i-n-e, you get the word "quiet," as in quieting the voices.

Seroquel, the brand name, shares the "q-u-e" from **quetiapine** and quell means to silence someone.

NEURO SECTION IX. ANTIEPILEPTICS

The traditional anti-epileptic drugs **carbamazepine, divalproex,** and **phenytoin** have been around for a long time and we usually know what to expect with their use.

We may have less experience with the newer anti-epileptics, such as **gabapentin** and **pregabalin,** but they are often just as effective as the traditional drugs.

Neurologists try medications in a patient until one drug relieves the seizure symptoms.

THREE TRADITIONAL ANTIEPILEPTICS

1) Carbamazepine (brand, Tegretol)
car-bah-MAZE-uh-peen (TEG-reh-tawl)

While **carbamazepine's** "-pine," p-i-n-e is the stem, it's not very useful because the "-pine" ending just means a chemical has three rings.

Instead, think either of seizures being "carbed" (curbed) or of being "amazed" that the seizures are "sto-pping" using the letters inside **carb-amaze-pine**.

I remember the brand name **Tegretol** has the scattered letters "t-r-o-l" in con<u>trol</u>, as in to control seizures; **Tegretol**.

2) Divalproex (brand, Depakote)
dye-VAL-pro-ex (DEP-uh-coat)

While you can find "v-a-l" in many medication names, it is helpful to think of the "val" in di<u>val</u>proex and it's similarity with the "vul, v-u-l" in con<u>vul</u>sions. I know it's a stretch.

One student thought of **dival-pro-ex** as a <u>pro</u> at <u>ex</u>tracting seizures; **divalproex**.

3) Pheny<u>toin</u> (brand, Dilantin)
FEN-eh-toyn (DYE-lan-tin)

The "toin, t-o-i-n" stem helps you remember **phenytoin** is an antiepileptic.

Dilantin and shakin' almost rhyme.

TWO NEWER ANTIEPILEPTICS

1) <u>Gab</u>apentin (brand, Neurontin)
GA-ba-PEN-tin (NER-on-tin)

The "gab, g-a-b" stem is a little misleading. Neither **gabapentin** nor **pregabalin** directly affect gamma-aminobutyric-acid (GABA) receptors. However, having them both have the same stem helps set a memory device for the newer antiepileptics.

The "neu, n-e-u" in **<u>Neu</u>rontin** is one way to remember that this is a newer drug.

2) Pre<u>gab</u>alin (brand, Lyrica)
pre-GAB-uh-lin (LEER-eh-ca)

A <u>lyre</u> is a musical instrument and a <u>lyric</u> is a line in a song.

Either way, you can think of a seizure coming back into harmony with **Lyrica**.

NEURO SECTION X. PARKINSON'S, ALZHEIMER'S AND MOTION SICKNESS

Many people associate Parkinson's disease with the celebrity actor Michael J. Fox of *Back to the Future* fame.

Using a combination drug like **levodopa** with **carbidopa** for Parkinson's works to restore dopamine, a neurotransmitter responsible for proper motor function that is seriously depleted by the disease.

In Alzheimer's, the issue is the neurotransmitter acetylcholine. An Alzheimer's drug like **donepezil** works to restore the neurotransmitter acetylcholine, by reducing its breakdown by acetylcholinesterase.

Concerning Alzheimer's, I remember vividly calling my grandmother on the telephone and telling her "I love you," to which she responded, "I hope your wife doesn't find out."

There is a cruelty in Alzheimer's, and the patients and their caregivers desperately need your help.

TWO PARKINSON'S MEDICATIONS

1) Levodopa with Carbidopa (brand, Sinemet)
LEE-vo-doe-pa / CAR-bid-oh-pa" (SIN-uh-met)
The stem dopa, d-o-p-a in both levodopa and carbidopa help remind us that increased dopamine is critical to dealing with the disease. The brand *Sinemet* combines **levodopa** and **carbidopa** to work *syn*ergistically.

Carbidopa doesn't actually have an antiparkinsonian effect, but it reduces the breakdown of **levodopa** so more is available to the patient from a smaller dose.

2) Selegiline (brand, Eldepryl)
se-LEDGE-eh-lean (EL-duh-pril)

The "-giline," g-i-l-i-n-e stem should be your hint that **selegiline** is a Parkinson's medication.

You can find the letters of the word senile in the generic name **selegiline**. I used to confuse this as an Alzheimer's medication, but it's not, so I was being a little senile.

The brand name **Eldepryl** helps you remember that it relieves symptoms of Parkinson's disease, a condition more prevalent in the "<u>elde</u>rly;" **Eldepryl**.

Two Alzheimer's Medications

I will use "memory is done," to remember *mem*antine and *don*epezil. That's why they are not alphabetically listed.

1) Memantine (brand, Namenda)
Meh-MAN-teen (Nuh-MEN-duh)

The generic name **memantine** has "mem, m-e-m" for <u>mem</u>ory in it.

The brand name **Namenda** comes from <u>N</u>-<u>M</u>ethyl-<u>D</u>-<u>a</u>spartate **(N-M-D-A)**, the receptor it antagonizes.

One of my students always thought of **Namenda** as <u>mend</u>ing the brain; **Namenda**.

2) Done<u>pezil</u> (brand, Aricept)
Doe-NEP-eh-zill (AIR-eh-sept)

When I think of **donepezil** as an Alzheimer's medication, I think, "I <u>don</u>'t remember <u>zil</u>ch!" taking the "d-o-n" from the front of donepezil and the "z-i-l" from the back. The brand name **Aricept** improves per<u>cep</u>tion and Alzheimer's patients' powers of recollection.

A student thought of the "air" in **Aricept** as helping a patient who acts somewhat spacey or air-headed.

A Motion Sickness Medication

At the beginning of this chapter, I mentioned the tablet form meclizine, brand name Dramamine, for motion sickness. There is also a patch that some people use on cruise ships.

Scopolamine (brand, Transderm-Scop)
sco-POL-uh-mean (trans-DERM SCOPE)

Transderm-Scop is a transdermal form of **scopolamine**, for motion sickness. "Trans" means across, "derm" means "skin," so across-the-skin **scopolamine**; **Transderm-Scop**

Chapter 5 Quiz: Neuro, Drug stem and drug classification practice

I'm going to read a generic name for you and I want you to think about the stem, if any, and the class of medication.

1. Benzocaine, caine, c-a-i-n-e, local anesthetic
2. Lidocaine, caine, c-a-i-n-e, local anesthetic
3. Meclizine, no stem, antivertigo
4. Acetaminophen P-M, (diphenhydramine), no stem, non-narcotic analgesic with sedating first generation antihistamine working as a sedative – hypnotic
5. Eszopiclone, clone, c-l-o-n-e, benzodiazepine-like sedative hypnotic
6. Zolpidem, pidem, p-i-d-e-m, benzodiazepine-like sedative hypnotic
7. Ramelteon, melteon, m-e-l-t-e-o-n, melatonin receptor agonist sedative hypnotic
8. Citalopram, no stem, Selective Serotonin Reuptake Inhibitor antidepressant, S-S-R-I
9. Escitalopram, no stem, Selective Serotonin Reuptake Inhibitor antidepressant, S-S-R-I
10. Sertraline, traline, t-r-a-l-i-n-e, Selective Serotonin Reuptake Inhibitor antidepressant, S-S-R-I

CHAPTER 5: NEURO

11. Fluoxetine, oxetine, o-x-e-t-i-n-e, Selective Serotonin Reuptake Inhibitor antidepressant, S-S-R-I
12. Paroxetine, oxetine, o-x-e-t-i-n-e, Selective Serotonin Reuptake Inhibitor antidepressant, S-S-R-I
13. Duloxetine, oxetine, o-x-e-t-i-n-e, careful, this is a Serotonin *Norepinephrine* Reuptake Inhibitor, S-N-R-I
14. Venlafaxine, faxine, f-a-x-i-n-e, Serotonin Norepinephrine Reuptake Inhibitor, S-N-R-I
15. Amitriptyline, triptyline, t-r-i-p-t-y-l-i-n-e, tricyclic antidepressant, T-C-A
16. Isocarboxazid, no stem, monoamine oxidase inhibitor antidepressant, M-A-O-I
17. Bupropion, no stem, atypical antidepressant and smoking cessation aid
18. Varenicline, nicline, n-i-c-l-i-n-e, smoking cessation aid
19. Alprazolam, azolam, a-z-o-l-a-m, benzodiazepine
20. Midazolam, azolam, a-z-o-l-a-m, benzodiazepine
21. Clonazepam, azepam, a-z-e-p-a-m, benzodiazepine
22. Lorazepam, azepam, a-z-e-p-a-m, benzodiazepine
23. Dexmethylphenidate, no stem, ADHD stimulant
24. Methylphenidate, no stem, ADHD stimulant
25. Atomoxetine, oxetine, o-x-e-t-i-n-e, careful, not an SSRI, but an ADHD non-stimulant
26. Lithium, no stem, simple salt and mood stabilizer
27. Chlorpromazine, no stem, first generation *low* potency antipsychotic
28. Haloperidol, peridol p-e-r-i-d-o-l, first generation *high* potency antipsychotic
29. Risperidone, peridone, p-e-r-i-d-o-n-e, 2nd generation antipsychotic
30. Quetiapine, tiapine, t-i-a-p-i-n-e, 2nd generation antipsychotic
31. Carbamazepine, pine, p-i-n-e, traditional antiepileptic
32. Divalproex, no stem, traditional antiepileptic
33. Phenytoin, toin, t-o-i-n, traditional antiepileptic

34. Gabapentin, gab, g-a-b, newer antiepileptic
35. Pregabalin, gab, g-a-b, newer antiepileptic
36. Levodopa with carbidopa, dopa, d-o-p-a, antiparkinson's
37. Selegiline, giline, g-i-l-i-n-e, antiparkinson's
38. Memantine, no stem, alzheimer's
39. Donepezil, pezil, p-e-z-i-l, alzheimer's
40. Scopolamine, no stem, motion sickness patch

NEURO: MEMORIZING THE CHAPTER

"N" NEURO

The four neuro OTC drugs are the local anesthetics **benzocaine** and **lidocaine**, which are an ester and amide respectively. **Meclizine** is for dizziness, and **acetaminophen** with **diphenhydramine** is for insomnia. Connect this with other anti-insomniacs the benzodiazepine-like sedative-hypnotics **eszopiclone**, **zolpidem**, and melatonin receptor agonist **ramelteon**. No sleep equals depression. Five SSRI antidepressants: **citalopram** and **escitalopram**, **sertraline** in the middle followed by three "–oxetines:" **fluoxetine** and **paroxetine** (the SSRIs) and **duloxetine** (the SNRI) and **venlafaxine** the SNRI; followed by, in order of safety, the TCA **amitriptyline** and the M-A-O-I **isocarboxazid**. Move from depression to smoking: **bupropion** and **varenicline**, smoking while anxious, two –azolams: **alprazolam** and **midazolam** and two -azepams **clonazepam** and **lorazepam**. From the "A" in anxious to the "A" in ADHD stimulants **dexmethylphenidate**, **methylphenidate**, to non-stimulant **atomoxetine**; to "m-o" from atomoxetine for m-o-o-d, mood stabilizer **lithium**; then "L" from lithium to low potency 1st-generation **chlorpromazine**, to high potency "halo" **haloperidol**, "–peridol" to "–peridone," 2nd generation whisper **risperidone** to whisper-quiet **quetiapine**; pine to traditional antiepileptics **carbamazepine**, **divalproex** and

phenytoin to two newer antiepileptics Neu-rontin (which is **gabapentin**) and **pregabalin**. From epileptic motion to Parkinsonian motion: **carbidopa with levodopa** and **selegiline** senility to Alzheimer's "memory is done" taking the m-e-m from **memantine** to d-o-n-e **donepezil** ending with "D" dizzy **scopolamine**.

Can you recite these in order?

Benzocaine	Amitriptyline	Chlorpromazine
Lidocaine		Haloperidol
	Isocarboxazid	
Meclizine		Risperidone
	Bupropion	Quetiapine
Acetaminophen PM	Varenicline	
		Carbamazepine
Eszopiclone	Alprazolam	Divalproex
Zolpidem	Midazolam	Phenytoin
Ramelteon		
	Clonazepam	Gabapentin
Citalopram	Lorazepam	Pregabalin
Escitalopram		
Sertraline	Dexmethylphenidate	Levodopa with
Fluoxetine	Methylphenidate	carbidopa
Paroxetine		Selegiline
	Atomoxetine	
Duloxetine		Memantine
Venlafaxine	Lithium	Donepezil
		Scopolamine

Once you're able to recite the list, you are ready to move on.

CHAPTER 6: CARDIO

CARDIO SECTION I. OTC
ANTIHYPERLIPIDEMICS AND ANTIPLATELET

Few over-the-counter (OTC) medications help a patient with cardiologic issues. Both **omega-3-acid ethyl esters** and **niacin** come as brand name and OTC products, and can reduce cholesterol's impact on the patient.

Plain **aspirin** in a low-dose of 81 milligrams helps prevent platelets from clotting, reducing a patient's chance of a heart attack.

TWO OTC ANTIHYPERLIPIDEMICS

1) Omega-3-acid ethyl esters (brand, Lovaza)
Oh-MEG-uh THREE AS-sid ETH-ill EST-ers (Loh-VAH-zah)

Omega-3-fatty acids are available over-the-counter, but there is also a prescription version that undergoes rigorous FDA testing.

Lovaza is a prescription brand name, but you can find **omega-3-fatty acids** over-the-counter commonly labeled as "Fish Oil."

2) Niacin (brand, Niaspan)
NYE-uh-sin (NYE-uh-span)

A vitamin like **niacin** can reduce cholesterol levels in the body, however it may cause facial flushing that an **aspirin** thirty minutes before treatment prevents.

This flushing unfortunately gets in the way of helping many patients who would otherwise benefit from an inexpensive cholesterol lowering medication.

OTC Antiplatelet

Aspirin (brand, Ecotrin)
AS-per-in (ECK-oh-trin)

The 81mg daily **aspirin** dosage is not for analgesia or fever reduction. Rather, it keeps platelets from sticking, helping prevent strokes and heart attacks.

CARDIO SECTION II. DIURETICS

There are five important structures in the nephron that I want to talk about in order. We begin with 1) the glomerulus, a bed of capillaries, to the 2) proximal convoluted tubule to 3) Loop of Henle to 4) distal convoluted tubule and 5) the collecting duct.

Something that is proximal is near or in close proximity. Something that is distant or distal is far.

In this case, we are talking about close to or far from the glomerulus.

The order of diuretics at these sites would then follow this order:

1. Osmotic diuretics like **mannitol** work at the proximal convoluted tubule (P-C-T).
2. Loop diuretics like **furosemide** affect the Loop of Henle.
3. Thiazide diuretics like **hydrochlorothiazide** work at the distal convoluted tubule (D-C-T).
4. Potassium sparing diuretics like **triamterene** or **spironolactone** work at the collecting duct.

Picture a water slide. A lot of water flows at the top (the glomerulus). A trickle flows at the bottom (the collecting duct).

Similarly, diuretics produce less diuresis as they continue down the waterslide. The order from most to least diuresis is osmotic > loop > thiazide > potassium sparing or translate that into drug names and it becomes mannitol, furosemide, hydrochlorothiazide, then triamterene and spironolactone.

OSMOTIC DIURETIC

Mannitol (brand, Osmitrol)
MAN-eh-tall (OZ-meh-trawl)

Mannitol, an osmotic diuretic reduces intracranial pressure in an emergency.

The brand name **Osmitrol** combines the class of medication "osmotic," and adds that it helps control brain swelling, taking the o-s-m-i from osmotic and t-r-o-l from control; **Osmitrol**.

The actor Bruce Lee died from this event.

LOOP DIURETIC

Furosemide (brand, Lasix)
Fyoor-OH-seh-mide (LAY-six)

Chemists named this class of diuretics after the part of the nephron the drug works in, the Loop of Henle.

While the "semide, s-e-m-i-d-e" stem indicates a "furosemide-type" diuretic, that's like defining a word with the word itself.

One student said, "I have to pee furiously" as her mnemonic since loop diuretics produce significant diuresis taking the f-u-and r from furosemide.

The brand name **Lasix** indicates it lasts six hours; **Lasix**.

THIAZIDE DIURETIC

Hydrochloro_thiazide_ (brand, Microzide)
High-droe-klor-oh-THIGH-uh-zide (MY-crow-zide)

Thiazide diuretics get their class name from the stem –thiazide, t-h-i-a-z-i-d-e of generic drugs like **hydrochlorothiazide**.

The abbreviation H-C-T-Z comes from "h" for h_y_dro, "c" for c_h_loro, "t" for _t_hia, and "z" for _z_ide.

Thiazides don't produce as much diuresis as loop diuretics, but are excellent for initial treatment of hypertension.

While the "hydro" in **hydrochlorothiazide** stands for the hy_d_rogen atom, you can think of "hydro" as "water" for diuretic.

The brand **Microzide** has thia_zide_'s last four letters, z-i-d-e, **Microzide**.

POTASSIUM SPARING WITH THIAZIDE DIURETIC COMBINATION

Triamterene with Hydrochloro_thiazide_ (brand, Dyazide)
try-AM-terr-een / High-droe-klor-oh-THIGH-uh-zide
(DIE-uh-zyde)

The combination of a potassium sparing diuretic (**triamterene**) and thiazide diuretic (**hydrochlorothiazide**) keeps potassium levels in balance while producing modest diuresis.

The brand name **Dyazide** takes the first two letters of triamterene's brand name **D_y_renium** plus the last five letters of **hydrochloroth_iazide_**, so d-y plus i-a-z-i-d-e; **Dyazide**.

POTASSIUM SPARING DIURETIC

Spironolactone (brand, Aldactone)
spear-oh-no-LACK-tone (Al-DAK-tone)

Spironolactone is another potassium sparing diuretic like triamterene, but this drug can cause gynecomastia.

Gynecomastia is an enlargement of male breasts. To remember **spironolactone** works in the collecting duct, I look at the "lactone," and think "last one," taking the l-a and o-n-e from lactone.

I know a lactone is a kind of chemical structure, but its place of action sticks in my head with this mnemonic.

To come up with the brand name, the manufacturer simply replaced the "spironol" of **spironolactone** with "ald, a-l-d" to make **Aldactone**.

The "ald" is especially important because **spironolactone** blocks _ald_osterone, an important steroid hormone that helps the body retain sodium and water when blood pressure drops.

ELECTROLYTE REPLENISHMENT

Potassium chloride (brand, K-Dur)
poe-TASS-ee-um klor-eyed (Kay-Dur)

Potassium chloride is a supplement often administered when a potassium sparing diuretic is contraindicated or when a loop diuretic lowers a patient's potassium levels.

The "K" in **K-Dur** is the chemical symbol for potassium.

The "Dur, d-u-r," is for long _dur_ation.

CARDIO SECTION III. UNDERSTANDING THE ALPHAS AND BETAS

We to break up the terms alpha adrenergic antagonist and beta adrenergic antagonist into three pieces: a) the alpha or beta, 2) the adrenergic, and 3) the antagonist.

Alpha or Beta

Alpha and beta are the first two letters of the Ancient Greek alphabet. It might be easier to think of the British pronunciation Bee-tah for Bay-tah that makes more sense alphabetically because the British pronounce the first syllable as a "B," the second letter in the Roman alphabet we use.

Confusion about *alpha*-adrenergic antagonists like **doxazosin** and *beta*-adrenergic antagonists like **propranolol** comes from seeing those receptor names, *alpha* and *beta*, instead of their drug classification – blood pressure lowering or antihypertensive pills.

Adrenergic

The prefix "adren" refers to the adrenal glands. The adrenal glands are *above* (ad) the *kidney* (renal) and secrete adrenaline. The suffix "ergic" refers to the Greek for "works like." Therefore, these drugs work like **adrenaline**. Note: **Adrenaline** and **epinephrine** are the same. **Epinephrine** uses the Greek translation of *above* (epi) and *kidney* (neph) to make **epinephrine** instead of the Latin form, **adrenaline**.

Antagonist

An adrenergic *agonist* works *like* adrenaline while an adrenergic *antagonist* works in the *opposite* way.

Let's review alpha adrenergic antagonist. Alpha plus adrenergic plus antagonist makes a drug that affects an alpha-receptor like the adrenal gland would and because it is an antagonist, it blocks the receptor; alpha adrenergic antagonist.

Doxazosin, the alpha-adrenergic antagonist, has multiple uses, including hypertension and benign prostatic hyperplasia (BPH). This is why classifying by the receptor name alpha makes more sense than pigeonholing its use as singular for blood pressure.

Beta Adrenergic Antagonists

Now, let's talk about the beta adrenergic antagonists.

By calling **propranolol, a beta-adrenergic antagonist,** an antihypertensive, we again pigeonhole a drug into one use. Beta-blockers are versatile and can treat angina pectoris, congestive heart failure, stage fright, and migraine, in addition to hypertension.

Furthermore, there are Beta-receptor sub-types. Beta-1 receptors concentrate in the heart (and we have one heart). Beta-2 receptors concentrate in the lungs (and we have two lungs). You can find them in other places in the body, but for our introductory purposes, it's useful to think in this way.

Beta Blocker Selectivity

1st generation

If a beta-blocker is *non-selective*, like the 1st generation **propranolol,** it can affect both beta-1 receptors in the heart to lower heart rate (that's good in a hypertensive patient), and block beta-2 receptors in the lungs and cause bronchoconstriction (that's bad in the asthmatic patient). As such, the asthmatic patient might have an adverse reaction to a drug that bronchoconstricts as a side effect.

2nd generation

We prefer the 2nd generation *selective* **metoprolol** to control blood pressure because it's selective for just the heart; it's classified as *beta-1 selective*. However, the body will try to compensate for this reduction in blood pressure by vasoconstricting arterioles.

3rd generation

Carvedilol, a 3rd-generation beta-blocker, shows that it might be the best choice because it has vasodilating effects to counteract the vasoconstriction as well as cardiac effects.

To tie this all together let me tell you about Edgar Allan Poe and his short story, "The Tell-Tale Heart."

Edgar Allan Poe, an American writer, wrote an essay called the Tell-Tale Heart around 1843. In the story, a man murders another and hides the body in the wall.

When the police come, he thinks he hears the man's heart and owns up to the crime. However, it's really his own tachycardic (rapid-beating) heart that he hears. If he had taken a beta-adrenergic antagonist that affected the beta-1 receptors in the heart, his heartbeat would not have been able to go so high and he could have comfortably spoken with the police.

Let's look at some specific examples of alpha and beta antagonists.

ALPHA-1 ANTAGONIST

Dox<u>azosin</u> (brand, C<u>ar</u>d<u>ur</u>a)
Docks-AS-oh-sin (car-DUR-uh)

Doxazosin's blockade of alpha-1 receptors causes vasodilation and subsequent reduction in blood pressure. Memorize the stem "azosin, a-z-o-s-i-n" as an alpha-blocker.

You can also link the brand name, as **Cardura** provides <u>du</u>rable <u>car</u>diac relief of hypertension, using the d-u-r-a from durable and c-a-r from cardiac to form **Cardura**.

CHAPTER 6: CARDIO

Alpha-2 Agonist

Clonidine (brand, Catapres)
KLAH-neh-deen (CAT-uh-press)

Clonidine works in the brain by affecting alpha-*two* receptors to reduce peripheral vascular resistance. You can look at the brand name **Catapres** and think of catabolize (break down) pressure (blood pressure) using the c-a-t-a from catabolize and p-r-e-s from pressure; **Catapres**.

Prescribers also use **clonidine** in A-D-H-D as single therapy or with stimulants like **methylphenidate,** brand name **Concerta**. I had the weirdest experience at the gym where a parent had a loud and lengthy discussion with a psychiatrist about her child's **clonidine** and **Concerta** while lifting weights, they were talking as loud as a rock *concert*.

I never forgot the **clonidine and Concerta** tandem after that.

Beta Blockers - 1ˢᵀ-Generation - Non-Beta-Selective

Propran<u>olol</u> (brand, Inder<u>al</u>)
Pro-PRAN-uh-lawl (IN-dur-all)

Propranolol's last four letters have the "–olol, o-l-o-l" beta-blocker stem. The "o-l-o-l" looks like two letter "b's" backwards to remember (b) beta (b) blocker.

Alternatively, if you "<u>o</u>h, <u>l</u>augh <u>o</u>ut <u>l</u>oud," using the o-l-o-l from those four words, you can think of your heart rate going down from stress relief.

If you think of the last "al," a-l in the brand name **Inder<u>al</u>** as it blocks "all" beta-receptors, you can remember this is a non-selective beta-adrenergic antagonist.

BETA BLOCKERS - 2ND-GENERATION - BETA-SELECTIVE

Aten_olol_ (brand, Tenormin)
uh-TEN-oh-lol (Teh-NOR-min)

While the "olol, o-l-o-l" in **atenolol** identifies this medication as a beta-blocker, a student *does* have to memorize that **atenolol** is 2nd generation.

She can do that by memorizing atenolol's position *after* a non-selective first generation **propranolol** in this book *or* by seeing the "ten" in **atenolol** and knowing it's divisible by *two*.

The "Ten, T-e-n" in the brand name **Tenormin** also matches to the "ten, t-e-n" in a**ten**olol to help connect the brand and generic names *Ten*ormin and a*ten*olol.

Metopr_olol_ tartrate (brand, Lopressor)
meh-TOE-pruh-lawl TAR-trait (low-PRESS-or)

Practitioners rarely highlight the distinction in salts like tartrate and succinate, but it's important to recognize, as **metoprolol** *tartrate* and **metoprolol** *succinate* work for different lengths of time.

It would have been nice if the succinate was short acting and the tartrate, long acting; then alphabetical order would have worked or "s" for short acting.

That it goes contrary to this logic is how I personally remember which is which.

You can use the brand name **Lopressor** to remind you that **Lopressor** l_o_wers blood p_ress_ure, using the l-o from lowers and p-r-e-s-s from pressure; Lopressor.

Metopr_olol_ succinate (brand, Toprol X-L)
meh-TOE-pruh-lawl SUCKS-sin-ate" (TOE-prall ex-ell)

Metoprolol succinate is a long-acting form of **metoprolol**. The X-L, often used to identify clothing as extra-large, indicates an extra-long acting effect in medications like **Toprol XL**.

BETA BLOCKERS - 3ʳᵈ-GENERATION - NON-BETA-SELECTIVE, VASODILATING

Carvedilol (brand, Coreg)
car-veh-DILL-awl (CO-reg)

I'm not sure if it was intentional to create a kind of hybrid stem with the "dil, d-i-l" replacing the first "o" in "olol, o-l-o-l," but you can remember **carvedilol** works by both vasodilation and beta-blockade in this way. The only official stem, however, is the "dil, d-i-l."

I remember the brand name **Coreg** because it regulates coronary function, using the r-e-g from regulates and c-o from coronary; **Coreg.**

CARDIO SECTION IV. THE RENIN-ANGIOTENSIN-ALDOSTERONE -SYSTEM DRUGS

The R-A-A-S, or renin-angiotensin-aldosterone system, controls blood pressure. By defining these four words, we can better understand how the drugs work.

1) Renin

The word **renin** comes from **renal** for kidneys, and this enzyme converts angiotensinogen to angiotensin I.

2) Angiotensin

Angiotensin converting enzyme (ACE) converts **angiotensin I** to **angiotensin II**. Angiotensin II is a potent vasoconstrictor and increases blood pressure when that is what our body needs.

3) Aldosterone

Aldosterone causes the retention of sodium and water, which can further increase blood pressure.

4) System

These components, the renin, angiotensin, and aldosterone all work together in a system. Renin-angiotensin-aldosterone-system, R-A-A-S.

Two major R-A-A-S drug classes

1) ACE Inhibitors (A-C-E-Is)

Angiotensin converting enzyme inhibitors (ACE inhibitors), with the stem "pril, p-r-i-l" as in **enalapril** and **lisinopril** stop the body from creating the potent vasoconstrictor angiotensin II, thereby reducing hypertension.

2) ARBs

ARBs, or angiotensin II receptor blockers, with the stem "sartan, s-a-r-t-a-n," such as **losartan, olmesartan,** and **valsartan,** block or inhibit the connection between angiotensin II and the receptor that would cause vasoconstriction. We often use ARBs as an alternative when a patient experiences a non-productive cough as a side effect from an ACE inhibitor.

Here is a mnemonic a student of mine who loved literature made up. It might help you remember the difference: D'artagnan the musketeer has to be "sartan, s-a-r-t-a-n" with the bARB of his blade, otherwise, he'll not be an ACE in April, p-r-i-l, I'm afraid.

TWO ANGIOTENSIN CONVERTING ENZYME INHIBITORS (ACEIs)

1) Enalapril (brand, Vasotec)
eh-NAL-uh-pril (VA-zo-teck)

Sometimes students simply refer to the ACE inhibitors like **enalapril** as "prils" based on the stem "pril, p-r-i-l." While patients take **enalapril** orally, **enalaprilat** with the ending p-r-i-l-a-t, is an injectable form and active metabolite of **enalapril**.

The brand **Vasotec** alludes to vasodilation on the vasculature, using the v-a-s in Vasotec.

2) Lisinopril (brand, Zestril)
lie-SIN-oh-pril (ZES-tril)

Lisinopril, like **enalapril**, has the "pril, p-r-i-l" stem and works to block the vasoconstricting effects of angiotensin II.

A student came up with "**Lisinopril** thrills an overworked heart, blocking angiotensin II from getting a start."

THREE ANGIOTENSIN II RECEPTOR BLOCKERS (ARBs)

1) Losartan (brand, Cozaar)
low-SAR-tan (CO-zar)

Learn angiotensin II receptor blockers by the suffix "sartan, s-a-r-t-a-n."

The brand name **Cozaar** looks like it has R-A-A-S backwards (for renin-angiotensin-aldosterone-system) with a "z" replacing the "s."

2) Olmesartan (brand, Benicar)
Ole-meh-SAR-tan (BEN-eh-car)

Olmesartan is another ARB identified by its "sartan, s-a-r-t-a-n" stem.

Memorizing Pharmacology

The brand name **Benicar** hints that the drug will be<u>ne</u>fit the <u>car</u>diac system, taking the b-e-n from benefit and c-a-r- from cardiac to make **Benicar**.

3) Valsartan (brand, Diovan)
val-SAR-tan (DYE-oh-van)

Identify the generic name **valsartan** by its "sartan, s-a-r-t-a-n" stem. **Diovan** has three of the letters of the generic name <u>v</u>als<u>a</u>rt<u>an</u>, the v-a-n to help connect the two.

CARDIO SECTION V. CALCIUM CHANNEL BLOCKERS (CCBs)

There are two major classes of calcium channel blocker and the seven and six syllable names might be a little intimidating. They are the non-die-high-dro-pyre-eh-deens and die-high-dro-pyre-eh-deens and form a simple distinction that helps you understand its mechanism of action.

You see, both calcium channel blocker classes, the non-dihydropyridines and dihydropyridines, are vasodilators opening up blood vessels.

The *non*-dihydropyridines **diltiazem** and **verapamil** also affect the heart directly and are antidysrhythmics that can put the heart back in proper rhythm.

Amlodipine and **nifedipine** are two dihydropyridines that only vasodilate.

If a patient needs a calcium channel blocker to prevent uterine contractions, **nifedipine** would be the best choice because it does not suppress the mother's and fetus's hearts as the non-dihydropyridines would.

In our three daughters' case, the doctor prescribed low dose **nifedipine** so the calcium channel blockers did not suppress four hearts – my wife's and three unborn daughters'.

Two Non-Dihydropyridines

1) Diltiazem (brand, Cardizem)
dill-TIE-uh-zem (CAR-deh-zem)

The "-tiazem, t-i-a-z-e-m" stem identifies **diltiazem** as a non-dihydropyridine.

The brand name **Cardizem** adds the first five letters from cardiac, c-a-r-d-i to the last three letters of the generic diltiazem, z-e-m, to make **Cardizem**.

2) Verapamil (brand, Calan)
ver-APP-uh-mill (KALE-en)

Verapamil has the "pamil, p-a-m-i-l" stem. One of my students came up with "Vera and Pam are ill and need this calcium blocking cardiac pill," adding the V-e-r-a from Vera and P-a-m from Pam to make **verapamil**.

Often we associate **verapamil** with constipation. My grandmother, a Navy nurse, used to put a **verapamil** tablet on my grandfather's breakfast cereal spoon.

I always thought my grandfather was silently praying before he ate. When I finally asked him why he was so quiet, he said something to the effect of, "I'm deciding whether I want to eat or poop today."

The brand name **Calan** takes c-a-l from the word calcium, and a-n from channel blocker, make C-a-l-a-n, **Calan**.

Two Dihydropyridines

1) Amlodipine (brand, Norvasc)
am-LOW-duh-peen (NOR-vasc)

Students usually recognize **amlodipine's** "dipine, d-i-p-i-n-e" stem, not only as a dihydropyridine, taking the d-i-p-i-n-e from the chemical class, but also as a dip, d-i-p, in, i-n blood pressure; **amlodipine**.

A way to remember the brand name **Norvasc** is to think of the "n-o-r" from <u>nor</u>malizes and the "v-a-s-c" from <u>vasc</u>ulature; **Norvasc**.

Nif<u>edipine</u> (brand, Procardia)
nigh-FED-eh-peen (pro-CARD-e-uh)

Nifedipine is another dihydropyridine with the "dipine, d-i-p-i-n-e" stem.

Procardia takes the "p-r-o" from "<u>pro</u>motes" and "c-a-r-d-i-a" from "<u>cardia</u>c" so you can remember the brand **Procardia** <u>pro</u>motes <u>cardia</u>c health.

CARDIO SECTION VI. VASODILATOR

<u>Nitro</u>glycerin (brand, Nitrostat)
nigh-trow-GLI-sir-in (NYE-trow-stat)

"Nitro-" is a World Health Organization (W-H-O) stem.

Nitroglycerin converts to nitric oxide, a vasodilator so make sure the patient sits when he takes the med because it causes significant dizziness.

With the brand **Nitrostat**, take the n-i-t-r-o from "<u>nitro</u>us" as in a street car's extra fuel and the concern that the patient and blood pressure drop quickly or "<u>stat</u>, s-t-a-t"; **Nitrostat**.

CARDIO SECTION VII. ANTIHYPERLIPIDEMICS

Medications for elevated cholesterol fall into several categories, including the "statin, s-t-a-t-i-n" class, which are more properly, called the H-M-G-Co-A reductase inhibitors, and the fibric acid derivatives.

It's better to recognize statins such as **atorvastatin** and **rosuvastatin** with the infix + suffix "vastatin, v-a-s-t-a-t-i-n" because it avoids the confusion with the generic name

nystatin, an antifungal medication that ends in "statin, s-t-a-t-i-n."

Two H-M-G-Co-A reductase inhibitors

1) Ator<u>vastatin</u> (brand, Lipitor)
uh-TORE-va-stat-in (LIP-eh-tore)

Start with the infix "va, v-a" and add s-t-a-t-i-n to get "vastatin, v-a-s-t-a-t-i-n" as a way to identify the H-M-G-Co-As.

Students use letters of the H-M-G-Co-A class to memorize potential adverse effects: "H" for <u>h</u>epatotoxicity, "M" for <u>m</u>yositis, "G" for gestation (that you can't use atorvastatin during pregnancy).

Brand name **Lipitor** is a <u>lip</u>id gladia<u>tor</u>, taking l-i-p from lipid and t-o-r from gladiator; **Lipitor**.

2) Rosu<u>vastatin</u> (brand, Crestor)
Row-sue-vuh-STAT-in (CRES-tore)

Like **atorvastatin, rosuvastatin** shares the "vastatin, v-a-s-t-a-t-i-n" ending. Remember **Crestor** de<u>cr</u>eases chol<u>ester</u>ol, taking the c-r from de<u>cr</u>eases and e-s-t-o-r from chol<u>ester</u>ol to make **Crestor**.

Fibric acid derivative

Feno<u>fibrate</u> (brand, TRICOR)
fen-oh-FIE-brate (TRY-core)

A drug like **fenofibrate** has the obvious stem "fibrate, f-i-b-r-a-t-e," a triglyceride lowering fibric acid derivative.

The brand name **TRICOR** takes the t-r-i from <u>tri</u>glycerides and c-o-r from <u>cor</u>onary; **TRICOR**.

Memorizing Pharmacology

CARDIO SECTION VIII. ANTICOAGULANTS AND ANTIPLATELETS

Four anticoagulants

Anticoagulants affect clotting factors to help prevent thrombosis. The injectable anticoagulants **enoxaparin** or **heparin** and the oral anticoagulant **warfarin** affect coagulation in slower moving blood vessels like veins. **Dabigatran** also works as an oral anticoagulant, but does not require monitoring with blood tests like **warfarin** and **heparin**.

Two antiplatelets

Platelets stop bleeding by creating clots. However, patients with excess cholesterol might have a plaque that makes the clot more likely to form in a dangerous place.

The antiplatelet drugs **aspirin** and **clopidogrel** decrease how "sticky" platelets are in high-pressure vessels such as arteries to prevent the clot and ensuing heart attack or stroke. We reviewed aspirin in the OTC section of this chapter.

LET'S LOOK AT THE FOUR ANTICOAGULANTS ENOXAPARIN, HEPARIN, COUMADIN, AND DABIGATRAN

1) Enoxa<u>parin</u> (brand, Lovenox)
e-knocks-uh-PEAR-in (LOW-ven-ox)

Enoxaparin and **heparin** share the "parin, p-a-r-i-n" stem because they are related. **Enoxaparin** is more expensive per dose, but patients can use it at home. It's also used as bridge therapy in a patient who is starting **warfarin** therapy.

Lovenox is a <u>lo</u>w molecular weight heparin for deep <u>vein</u> thrombosis prevention, taking the l-o from low and v-e-n from vein, **Lovenox**.

2) Heparin
HEP-uh-rin

Heparin and "bleedin'" sort of rhyme to remember its primary adverse effect. A student mentioned the actor Dennis Quaid's twins, who received a double dose of **heparin** that caused bleeding. Sometimes knowing a celebrity who has delt with a condition helps place the drug in memory.

3) War<u>farin</u> (brand, Coumadin)
WAR-fa-rin (KOO-ma-din)

That the "parin, p-a-r-i-n" stem from the anticoagulants **heparin** and **enoxaparin** and "farin, f-a-r-i-n" stem of **warfarin** are similar. This reminds students they are all anticoagulants.

Students associate bleeding with <u>warfar</u>e and warfarin has the w-a-r-f-a-r from warfare to recognize bleeding is a potential adverse effect.

The <u>I</u>-<u>N</u>-<u>R</u> (<u>i</u>nternational <u>n</u>ormalized <u>r</u>atio), a way of measuring **warfarin's** effectiveness, monitors the patient who is on therapy.

"I-N-R" happen to be the last three letters of warfa<u>rin</u> to help you remember.

A student said that war<u>far</u>in has "far, f-a-r" in it, as in you have to go far to have blood drawn.

A way to remember Vitamin K affects warfarin's brand name **Coumadin** and coagulation is to spell **Coumadin** with a "K" instead of a "C." K-o-u-m-a-d-i-n.

4) Dabi<u>gatran</u> (brand, Pradaxa)
da-bih-GA-tran (pra-DAX-uh)

Memorize **dabigatran's** "gatran, g-a-t-r-a-n" stem to note the difference between anticoagulants.

Dabigatran doesn't need INR monitoring like warfarin does and you can note the last three letters in dabigatran as not being "I-N-R."

ANTIPLATELET

Clopidogrel (brand, Plavix)
klo-PID-oh-grel (PLA-vix)

Clopidogrel has the "grel, g-r-e-l" stem identifying its class.

Clopidogrel and **aspirin** work similarly leading to a reduced likelihood that platelets will stick together and clot.

Plavix vexes platelets taking the "v" and "x" from vexes and p-l-a from platelets to keep the blood thin; **Plavix**.

CARDIO SECTION IX. CARDIAC GLYCOSIDE AND ANTICHOLINERGIC

A cardiac glycoside, such as **digoxin**, increases the force of contraction of the heart. We call this a positive inotropic effect. Also an antidysrhythmic, **digoxin** changes the electrochemistry of the heart to prevent dysrhythmias.

Atropine, an anticholinergic, prevents bradycardia, a drop in heart rate. **Atropine** can treat certain cholinergic poisonings as well.

CARDIAC GLYCOSIDE

Digoxin (brand, Lanoxin)
di-JOCKS-in (la-KNOCKS-in)

Digoxin treats congestive heart failure by increasing the force of the heart's contractions.

Digoxin is derived from the plant *Digitalis lanata*. In Latin, *Digitalis* means something like hand or "digits," while *lanata* means "wooly" because the actual plant is fuzzy.

Therefore, **digoxin** takes the d-i-g from *digitalis*, and the brand name **Lanoxin** takes the l-a-n from *lanata*.

Alternatively, you could remember that **Lanoxin** and **digoxin** keep your heartbeat r*o*ck*in*'.

ANTICHOLINERGIC INJECTION

A<u>trop</u>ine (brand, AtroPen)
ah-trow-PEEN (ah-trow-PEN)

Atropine has the –tropine, t-r-o-p-i-n-e stem.

Atropine causes anticholinergic (anti = against, cholinergic = of acetylcholine) effects.

Anticholinergic effects fall under the broad category of "dry."

Use the ABDUCT, capital A-B-D-U-C-T mnemonic, as in anticholinergics "abduct" water to remember these six effects.

<u>A</u>nhidrosis (A), <u>B</u>lurry vision (B) (secondary to dry eyes), <u>D</u>ry mouth (D), <u>U</u>rinary retention (U), <u>C</u>onstipation (C), and <u>T</u>achycardia (T).

This tachycardic side effect therapeutically prevents bradycardia in patients undergoing certain procedures.

Note the opposite: Cholinergic effects would include "wet" effects: sweating, lacrimation (watery eyes), hypersalivation, urinary incontinence, diarrhea, and bradycardia.

Chapter 6 Quiz: Cardio, Drug stem and drug classification practice

I'm going to read a generic name for you and I want you to think about the stem, if any, and the class of medication.

1. Omega-3-Acid Ethyl Esters, no stem, antihyperlipidemic
2. Niacin, no stem, vitamin B-3, antihyperlipidemic
3. Aspirin (Low Dose), no stem, over-the-counter anti-platelet
4. Mannitol, no stem, osmotic diuretic
5. Furosemide, semide, s-e-m-i-d-e, loop diuretic
6. Hydrochlorothiazide, thiazide, t-h-i-a-z-i-d-e, thiazide diuretic
7. Hydrochlorothiazide with Triamterene, thiazide, t-h-i-a-z-i-d-e, thiazide diuretic with triamterene, no stem, potassium sparing diuretic
8. Spironolactone, no stem, potassium sparing diuretic
9. Potassium Chloride, no stem, electrolyte replenishment
10. Doxazosin, azosin, a-z-o-s-i-n, alpha 1 antagonist for high blood pressure
11. Clonidine, no stem, alpha 2 agonist
12. Propranolol, olol, o-l-o-l, 1st generation non beta selective beta blocker
13. Atenolol, olol, o-l-o-l, 2nd generation, beta one selective beta blocker
14. Metoprolol Tartrate, olol, o-l-o-l, 2nd generation, short acting beta one selective beta blocker
15. Metoprolol Succinate, olol, o-l-o-l, 2nd generation, long acting beta one selective beta blocker
16. Carvedilol, dil, d-i-l, vasodilating 3rd generation non-selective beta blocker; note d-i-l-o-l is an offshoot of –o-l-o-l

CHAPTER 6: CARDIO

17. Enalapril, pril, p-r-i-l, angiotensin converting enzyme, ACE, inhibitor
18. Lisinopril, pril, p-r-i-l, angiotensin converting enzyme, ACE, inhibitor
19. Losartan, sartan, s-a-r-t-a-n, angiotensin II receptor blocker, ARB
20. Olmesartan, sartan, s-a-r-t-a-n, angiotensin II receptor blocker, ARB
21. Valsartan, sartan, s-a-r-t-a-n, angiotensin II receptor blocker, ARB
22. Diltiazem, tiazem, t-i-a-z-e-m, non-dihydropyridine calcium channel blocker (C-C-B)
23. Verapamil, pamil, p-a-m-i-l, non-dihydropyridine calcium channel blocker (C-C-B)
24. Amlodipine, dipine, d-i-p-i-n-e, dihydropyridine calcium channel blocker (C-C-B)
25. Nifedipine, dihydropyridine calcium channel blocker (C-C-B)
26. Nitroglycerin, nitro, n-i-t-r-o, vasodilator
27. Atorvastatin, infix "va" + stem "statin," vastatin, v-a-s-t-a-t-i-n, H-M-G Co-A reductase inhibitor, a type of antihyperlipidemic
28. Rosuvastatin, infix "va" + stem "statin," vastatin, v-a-s-t-a-t-i-n, H-M-G Co-A reductase inhibitor, a type of antihyperlipidemic
29. Fenofibrate, fibrate, f-i-b-r-a-t-e, fibric acid derivative antihyperlipidemic
30. Heparin, parin, p-a-r-i-n, parenteral anticoagulant
31. Enoxaparin, parin, p-a-r-i-n, parenteral low molecular weight heparin
32. Warfarin, farin, f-a-r-i-n, oral anticoagulant
33. Dabigatran, gatran, g-a-t-r-a-n, oral anticoagulant
34. Clopidogrel, grel, g-r-e-l, oral antiplatelet
35. Digoxin, no stem, cardiac glycoside
36. Atropine, tropine, t-r-o-p-i-n-e, anticholinergic

CARDIO: MEMORIZING THE CHAPTER

"C" CARDIO

"O" from cardio to **Omega-3 acid ethyl esters** to **niacin** to **aspirin** you need to take before the niacin to prevent flushing. Flushing five diuretics in nephron order: **mannitol, furosemide, hydrochlorothiazide, hydrochlorothiazide with triamterene, spironolactone** (potassium sparing), then **potassium chloride** to alphas: alpha 1 **doxazosin**, alpha 2 **clonidine**; to betas: 1st-generation beta 1 and beta 2 non-selective **propranolol**; 2nd-generation, beta 1 only, **atenolol, metoprolol** tartrate short acting to **metoprolol** succinate long acting; to 3rd-generation **carvedilol**. Dil to pril ACE inhibitors **enalapril, lisinopril**. ARBs with letters "love," "l-o-v;" **losartan, olmesartan** and **valsartan**; from "-sartan" to CCBs non-dihydropyridines **diltiazem** and **verapamil** to the dihydropyridines **amlodipine** and **nifedipine** vasodilating only; CCBs to **nitroglycerin**, a vasodilator. Nitroglycerin brand name **Nitrostat** to stat-ins: **atorvastatin** to **rosuvastatin**; LDL to lowered triglycerides, **fenofibrate** - fibs children tell to parents, parenteral **enoxaparin** and **heparin**, "-parin" to "-farin," **warfarin** enteral anticoagulant with **dabigatran**, "d" to **digoxin** for CHF and **atropine** to prevent a bradycardic mess.

CHAPTER 6: CARDIO

Can you recite these in order?:

Omega-3-Acid Ethyl Esters	Propran_olol_	Ni_tro_glycerin
Niacin	Aten_olol_	
	Metop_rolol_ Succinate	A_torvastatin_
	Metop_rolol_ Tartrate	Rosu_vastatin_
Aspirin (Low Dose)	Carve_dilol_	
		Feno_fibrate_
Mannitol	Enala_pril_	
Furo_semide_	Lisino_pril_	He_parin_
Hydrochloro_thiazide_		Enoxa_parin_
Hydrochlorothiazide with triamterene	Los_artan_	
	Olme_sartan_	War_farin_
Spironolactone	Val_sartan_	Dabi_gatran_
Potassium Chloride	Dil_tiazem_	Clopido_grel_
	Vera_pamil_	
Dox_azosin_		Digoxin
Clonidine	Amlo_dipine_	
	Nife_dipine_	Atropine

Before you move on from this list, I have an additional mnemonic song for you.

CARDIODE TO JOY

Cardiode to Joy, sung to Beethoven's *Ode to Joy*, is a mnemonic that you can sing, hum or just say. It connects common cardiologic drug suffixes, pharmacologic classes and therapeutic roles. *Cardiode to Joy* is a bonus track you can memorize while you're doing Cardio. Here goes:

o-l-o-l-p-r-i-l-and-s-a-r-t-a-n

be-ta-block-er-ace-in-hib-i-tor-and-ARBs-suff-ix-end

as-pir-in-and-clo-pid-o-grel-both-block-plate-lets-round-a-stent

war-fa-rin-and-hep-a-rin-are-both-an-ti-co-ag-u-lants

stat-ins-low-er-chol-est-ter-ol

dig-keeps-your-heart-from-fail-in

ver-a-pa-mil-and-am-lo-di-pine

both-block-cal-cium-chan-nels.

A quick *Cardiode to Joy* explanation

The first line's stems, the O-l-o-l-p-r-i-l-and-s-a-r-t-a-n represent the suffixes for the second line's therapeutic classes. These include beta-blockers, Angiotensin Converting Enzyme Inhibitors, and Angiotensin II Receptor Blockers; olol, pril, and sartan, respectively

Song line three pairs aspirin and clopidogrel, two antiplatelets and line four pairs warfarin, and heparin, two anticoagulants.

I separated them as a reminder that antiplatelets and anticoagulants have different mechanisms of action.

Lines five and six outline the therapeutic benefits of statins for cholesterol lowering and digoxin, abbreviated "dig," for congestive heart failure.

The last lines pair a non-dihydropyridine calcium channel blocker (verapamil) and a dihydropyridine calcium channel blocker (amlodipine) and the obvious mechanism of action to block calcium channels.

CHAPTER 7: ENDOCRINE AND MISCELLANEOUS

ENDOCRINE AND MISCELLANEOUS SECTION I. OTC INSULIN AND EMERGENCY CONTRACEPTION

REGULAR INSULIN AND N-P-H INSULIN

Most people don't think of insulin as an over-the-counter medication. Self-pay patients can purchase **regular insulin** or **N-P-H insulin** with an intermediate duration of action without a prescription.

Note: Alphabetically "N" comes before "R," but the convention is to put insulins in order from shortest to longest acting.

While a person can get insulin over-the-counter, I want to tell you a story about a miscommunication that resulted in some unfortunate temporary suffering for a dog.

THE INSULIN ORANGE

An owner and his dog came into the veterinarian's practice. The dog's condition needed insulin injections. The owner, a non-diabetic, had never used insulin before and needed some help.

The veterinarian allowed the patient to practice injecting an orange. Satisfied the owner's skill level was up to par, the vet let the two go on their way.

A week later, the dog returned with the same condition. Again, the veterinarian watched the owner inject the orange. The veterinarian said, "I don't know what the problem is." The owner replied, "Well, I don't think it's the injections that are the problem, it's just the dog won't eat the orange."

Even if the dog ate the orange, currently insulin isn't effective in p-o (oral) form. This reminds us that second conversations are okay. Sometimes we need to adjust to provide proper care.

Insulins are expensive, so drugstores keep them in the pharmacy refrigerator. This prevents theft and insures the insulin hasn't left the refrigerated temperatures. An insulin vial's box has the rectangular shape of a refrigerator to help you remember pharmacies refrigerate insulins for stability.

OVER THE COUNTER EMERGENCY CONTRACEPTION

Emergency contraception is a form of birth control used after unprotected sex to keep a patient from getting pregnant. **Levonorgestrel** as brand name **Plan B** is a recent introduction. In 1999, Plan B required a prescription. Later, patients could buy it from behind-the-counter (B-T-C). Now, it's available O-T-C.

Let's take a closer look at the names of the two over-the-counter insulins.

Regular insulin (brand, Humulin R)
REG-you-lar IN-su-lin (HUE-myou-lin ARE)

Regular insulin is short acting. Don't confuse this insulin with the shortest-acting insulins available, such as insulin lispro, brand **Humalog**. Prescribers give **regular insulin** when patients need to adjust dosages on a sliding scale.

Insulin used to come from a pig (porcine) or cow (bovine), but now matches human insulin because of molecular engineering.

CHAPTER 7: ENDOCRINE and MISCELLANEOUS

Therefore, the Eli Lilly brand name **Humulin** simply squishes the words <u>hum</u>an and ins<u>ulin</u> together, taking the h-u-m from human and u-l-i-n from insulin; **Humulin**.

NPH insulin (brand, Humulin N)
en-pee-aitch IN-su-lin (HUE-myou-lin EN)

The "N-P-H" in **NPH insulin** stands for <u>n</u>eutral <u>p</u>rotamine <u>H</u>agedorn. The "neutral protamine" refers to how Hagedorn, the inventor, chemically altered the insulin.

The "e-n" pronunciation of the letter "N" sounds a little like "i-n" and can help you remember it's an <u>in</u>termediate acting <u>in</u>sulin, taking the i-n from both intermediate and insulin; **Humulin N**.

OVER THE COUNTER EMERGENCY CONTRACEPTION

Levonor<u>gest</u>rel (brand, Plan B One-Step)
LE- vo- nor-JESS-trel (plan-bee won-step)

You can recognize **levonorgestrel** as a progestin hormone product by the "gest, g-e-s-t" stem. Take it within 72 hours after sexual intercourse. It causes nausea, but flat soda can calm this down. It's now called **Plan B One-Step** because it used to take two steps or two doses to provide this contraception.

I used to work in a college town pharmacy. Every Saturday and Sunday morning, I would have many students come to pick up **Plan B**. Every time, the man drove, and the woman sat in the passenger seat. When I told the male student, it was fifty bucks for the Plan B, he would invariably look at her, and then pay. One time, however, I overheard her say, "Oh no, you *just* didn't." from that "just, j-u-s-t" remember the "gest, g-e-s-t" that is part of **levonor<u>gest</u>rel**.

ENDOCRINE AND MISCELLANEOUS
SECTION II. PRESCRIPTION ONLY DIABETES MEDICATIONS

What is diabetes? Diabetes mellitus is a condition of chronic excess blood sugar. There are three types: type I, which we previously called juvenile onset diabetes; type II, which we referred to as adult-onset diabetes; and gestational diabetes, a condition where a pregnant woman *becomes* diabetic.

Depending on the condition, there are different drugs that can help lower blood sugar. Almost all oral medications have "g-l," or "g-l-u" for "<u>glu</u>cose" in their generic and sometimes brand names.

Four of these oral drugs include **metformin, brand <u>Glu</u>cophage** and the three generic drugs **sita<u>gl</u>iptin, <u>gl</u>ipizide, and <u>gl</u>yburide.**

I memorized them in alphabetical order of their drug classes: (b) biguanide (**metformin**), (d) dipeptidyl peptidase-4 (D-P-P-4) (**sitagliptin**), and (s) sulfonylurea (**glipizide** and **glyburide**).

Insulin for diabetes comes from the Latin word insula, which means island. The islets or islands of Langerhans in your pancreas have cells that produce insulin (beta cells), which lowers blood sugar; and cells that produce glucagon (alpha cells), which tells the body to raise blood glucose levels.

There are four major classes of insulin used in treatment:

- *Rapid acting* starts working in 15 minutes and lasts 3 to 5 hours, for example, **insulin lispro (brand, Humalog)**.
- *Short acting* works in 30-45 minutes and lasts up to 12 hours, for example, **regular insulin (brand, Humulin R)**.

CHAPTER 7: ENDOCRINE and MISCELLANEOUS

- *Intermediate duration* **N-P-H insulin (brand, Humulin N)** has an onset around 3 hours and works for about 18 hours.
- *Long duration* **insulin glargine (brand names Lantus and Toujeo)** starts working within 3 hours and it lasts up to 24 hours.

ORAL ANTI-DIABETICS - BIGUANIDE

Metformin (brand, Glucophage)
met-FOUR-men (GLUE-co-fage)

Use the "formin, f-o-r-m-i-n" stem to remember **metformin** is a biguanide anti-diabetic.

One student came up with this mnemonic, "If you met four men on **Glucophage**, they are diabetic then."

Phagocytosis is the process of cell eating. You can use the brand name **Glucophage**, G-l-u-c-o-p-h-a-g-e to think of the medication as eating, taking the g-l-u-c-o from glucose and p-h-a-g-e from phage-ing; **Glucophage**.

ORAL ANTI-DIABETICS- DPP-4 INHIBITOR

Sitagliptin (brand, Januvia)
sit-uh-GLIP-tin (ja-NEW-vee-uh)

Although the "gliptin, g-l-i-p-t-i-n" stem helps us recognize **sitagliptin** as an anti-diabetic, students associate the sugar you might put in "Lipton, L-i-p-t-o-n" iced tea with **sitagliptin**.

The brand name **Januvia** ends in "via, v-i-a" and is similar to the "dia, d-i-a" from diabetes.

Memorizing Pharmacology

ORAL ANTI-DIABETICS - 2ND-GENERATION SULFONYLUREAS

Glipizide (brand, Glucotrol)
GLIP-eh-zide (GLUE-co-trawl)

Your first question should be, "what happened to the first generation sulfonylureas?" Sometimes, when a medication's next generation so surpasses the previous as the preferred product, we simply don't include them anymore. First generation sulfonylureas cause more effects that are adverse. There is little reason to use them anymore.

The "gli-, g-l-i" stem in **glipizide** indicates an antihyperglycemic medication. The brand name **Glucotrol** alludes to the glucose control in diabetics taking the g-l-u-c from glucose and t-r-o-l from control; **Glucotrol**.

Glyburide (brand, DiaBeta)
GLY-byour-ide (die-uh-BAY-ta)

The "gly, g-l-y" stem in **glyburide** is archaic. The stem "gli, g-l-i," replaces it in new sulfonylurea antidiabetics.

The brand name **DiaBeta** combines the "dia, d-i-a" from diabetic, and the "B-e-t-a" from the Beta cells, which release insulin; **DiaBeta**.

While diabetes is an issue with excess glucose, hypoglycemia, too little glucose, also requires treatment.

HYPOGLYCEMIA TREATMENT

Glucagon (brand, GlucaGen)
GLUE-ca-gone (glue-ca-JEN)

Remember the generic name glucagon with, "I use **glucagon** when the glucose is gone," taking the g-l-u-c from glucose and g-o-n from gone; **Glucagon**.

CHAPTER 7: ENDOCRINE and MISCELLANEOUS

GlucaGen, the brand name, is a <u>gluc</u>ose <u>gen</u>erator taking the g-l-u-c from glucose and g-e-n from generator; **GlucaGen.**

Two Prescription-only Insulins, insulin lispro, ultra-short acting and insulin glargine, long acting

1) Insulin lispro (brand, Humalog)
IN-su-lin LICE-pro (HUE-mah-log)

Insulin glargine would precede **insulin lispro** alphabetically, but the convention is to list the medications shorter acting to longer acting. The complete order from shortest to longest acting is 1) insulin lispro, 2) regular insulin, 3) insulin NPH, and 4) insulin glargine.

I use my Spanish language to help memorize that **insulin lispro** is rapid acting. When I was in Mexico on a zip line, the person in the first tower would say, "Listo, listo," meaning "You ready, I'm ready." Then I would fly fast down that zip cord. I just replace the lispro's "s-p," with "t" because you must be "listo, listo" to take this rapid acting insulin with a meal.

Insulin lispro's brand name, **Humalog,** is a <u>hum</u>an insulin ana<u>log</u> combining the h-u-m from human and l-o-g from analog. I always pictured a <u>hum</u>an on a <u>log</u> floating down a rapids river to remember **Humalog** as a rapid acting insulin.

2) Insulin glargine (brand names Lantus and Toujeo)
IN-su-lin GLAR-Jean (LAN-tuss, TWO-jzeh-oh)

With **insulin glargine**, I think of the g-l-a-r in <u>glar</u>ing and i-n of lurking, someone who is *slowly* creeping around.

For the brand **Lantus**, one student came up with "**Lantus** la**sts** all day long, take it at night, and your life will be prolonged." I've also heard "**l**a**z**y **Lantus**" used to help remember it's a 24-hour drug.

ENDOCRINE AND MISCELLANEOUS SECTION III. THYROID HORMONES

Thyroid hormone stimulates the heart, metabolism, and helps with growth. A *hyper*thyroid patient's body uses energy too quickly because of the extra thyroid hormone in circulation.

This patient can use **propylthiouracil** to reduce the effects of the thyroid hormone. *Hypo*thyroid patients need extra thyroid hormone, such as **levothyroxine,** for replacement.

*H*YPOTHYROIDISM

Levothyroxine (brand, Synthroid)
Lee-vo-thigh-ROCKS-een (SIN-throyd)

The generic name **levothyroxine** has the "thyro, t-h-y-r-o" from thyroid in the name. You just have to remember it's for supplementation.

The brand name **Synthroid** combines the words syn<u>th</u>etic and thy<u>roid</u>, taking the s-y-n-t-h from synthetic and r-o-i-d from thyroid; **Synthroid**.

*H*YPERTHYROIDISM

Propylthio<u>u</u>racil (brand, P-T-U)
pro-pill-thigh-oh-YOUR-uh-sill (pee-tee-you)

PTU takes "p" from **p**ropyl, "t" from "**t**hio," and "u" from **u**racil in the generic **p**ropyl<u>t</u>hio<u>u</u>racil.

Although "thio, t-h-i-o" suggests a sulfur atom, you can think of it as <u>thy</u>roid <u>l</u>owering, using t-h-i from thyroid and "o" from lowering; propyl*thio*uracil.

ENDOCRINE AND MISCELLANEOUS SECTION IV. HORMONES AND CONTRACEPTION

Testosterone is an androgen steroid hormone that naturally comes from the male testes. As a medication, prescribers use it to supplement conditions of low testosterone.

Pharmaceutical birth control, commonly known as "the pill," traditionally came from a combined oral contraceptive pill (C-O-C-P) that has a combination of an estrogen and a progestin.

There are many variations of "the pill" including **Loestrin 24 F-e.** The F-e stands for <u>fe</u>rrous, or iron, on the Periodic Table of Elements.

The tri-phasic birth controls, such as **Tri-Sprintec**, have three different doses of an estrogen and progestin, taken throughout the month to mimic the body's naturally changing hormone levels.

Two novel birth control delivery methods include a vaginally inserted ring **(brand, NuvaRing)** and a transdermal patch **(brand, OrthoEvra)**.

TESTOSTERONE REPLACEMENT

Testo<u>ster</u>one (brand, AndroGel)
Tess-TOSS-ter-own (ANN-droh GEL)

Most people know the steroid hormone **testosterone**, but note the stem for a steroid is "ster, s-t-e-r" "<u>Andro</u>, A-n-d-r-o" is the Greek prefix for male and gel is the vehicle in **AndroGel**, indicating it's a "male's gel."

CONTRACEPTION - 2 COMBINED ORAL CONTRACEPTIVES

1) Noreth<u>in</u>drone with ethinyl <u>estr</u>adiol and ferrous fumarate (brand, Loestrin 24 Fe)
Nor-eth-IN-drone / ETH-in-ill es-tra-DYE-all
(Low-ES-trin EF-ee Twen-TEE fore)

It's important to first memorize the estrogen stem "estr, e-s-t-r" and progestin stem "gest, g-e-s-t."

To remember the brand, use this mnemonic: **Loestrin 24 Fe** reduces the length of menstruation, with iron supplementation, to prevent an anemic situation."

2) Norg<u>est</u>imate with ethinyl <u>estr</u>adiol (brand, Tri-Sprintec)
Nor-JESS-teh-mate / ETH-in-ill es-tra-DYE-all"

Again, look to the estrogen stem "estr, e-s-t-r" and progestin stem "gest, g-e-s-t."

Here's another mnemonic: "**Tri-Sprintec** is <u>tri</u>phasic, take three different doses, in seven-day spaces."

CONTRACEPTION - PATCH

Norelgestromin with ethinyl <u>estr</u>adiol (brand, OrthoEvra)
Nor-el-JESS-tro-min / ETH-in-ill es-tra-DYE-all
(OR-thoe EV-rah)

One student associated **norelgestromin's** "Norel" as "<u>not</u> ora<u>l</u>" to remember this is a patch. Another mnemonic highlights the drug's correct placement and length of therapy.

"**OrthoEvra** is a patch, put it on your arm, your abs, your buttock or back, and then take it off a week after that."

CHAPTER 7: ENDOCRINE and MISCELLANEOUS

CONTRACEPTION - RING

Etono<u>gest</u>rel with ethinyl <u>est</u>radiol (brand, NuvaRing)
Et-oh-no-JESS-trel / ETH-in-ill es-tra-DYE-all"
(NEW-va-ring)

A student thought of **etonogestrel**'s "Etono" in as "Eat, oh no" to remember it's not an oral tablet.

The brand **NuvaRing** combines "n-u" from new, "v-a" from vaginal, and "r-i-n-g" from "ring;" **Nuvaring**.

ENDOCRINE AND MISCELLANEOUS SECTION V. OVERACTIVE BLADDER, URINARY RETENTION, ERECTILE DYSFUNCTION, BENIGN PROSTATIC HYPERPLASIA

Frequently patients confuse the terms overactive bladder, urinary retention, erectile dysfunction, and benign prostatic hyperplasia:

- **Overactive bladder (abbreviated O-A-B) (incontinence)** is an inability to retain urine due to an overactive bladder.
- **Urinary retention** is a difficulty in urination.
- **Erectile dysfunction (abbreviated E-D) (impotence)** is the inability to achieve or maintain an erection.
- **Benign prostatic hyperplasia (abbreviated B-P-H)** is a benign (not harmful) prostate growth or increase in size.

Avoid the harsh terms "incontinence" or "impotence." Favor overactive bladder and erectile dysfunction instead.

THREE OVERACTIVE BLADDER MEDICATIONS

1) Oxybutynin (brand names Ditropan and Oxytrol O-T-C)
ox-e-BYOU-tin-in (DIH-trow-pan OX-ee-trawl OTC)

One student remembered **oxybutynin** as keepin' the urin' in.

Both **Di*tro*pan** and **Oxy*tro*l** have the "t-r-o" from control for cont*ro*lling an overactive bladder.

2) Solifenacin (brand, VESIcare)
sol-eh-FEN-a-sin (VEH-si-care)

Solifenacin's stem, "fenacin, f-e-n-a-c-i-n," should be your first clue to its function.

We give **solifenacin** once daily, so thinking about it as "slow-fenacin" helps in remembering this point.

In addition, **solifenacin** solves the problem of urine that needs to be "fenced in," using the "f-e-n-c" from fenced and "i-n" to almost complete the stem "fenacin."

The brand **VESIcare** contains *vesica*, v-e-s-i-c-a, which means "bladder" in Latin, **VESIcare**.

3) Tolterodine (brand, Detrol)
toll-TER-oh-dean (DEH-trawl)

The generic name **tolterodine** has the "t-r-o" from cont*ro*l in the name as well.

The brand **Detrol** helps detrusor muscle cont*ro*l, keeping urine in, taking the d-e-t-r from detrusor and –r-o-l from control; **Detrol**.

A URINARY RETENTION MEDICATION

Bethanechol (brand, Urecholine)
beh-THAN-uh-call (yur-eh-CO-lean)

The "chol, c-h-o-l" in **bethanechol** helps you remember it's *chol*inergic.

CHAPTER 7: ENDOCRINE and MISCELLANEOUS

While *anti*cholinergics are dry, cholinergics do the opposite and make things wet.

This drug assists the bladder muscles in expelling urine for a patient with urinary retention.

The brand name **Urecholine** alludes to how it affects urination through cholinergic effects, taking the u-r from urination and c-h-o-l from cholinergic; **Urecholine**.

ERECTILE DYSFUNCTION (TWO P-D-E-5 INHIBITORS)

1) Sildenafil (brand, Viagra)
sill-DEN-uh-fill (vie-AG-rah)

The –afil, a-f-i-l stem specifies the P-D-E-5 inhibitor class.

There is a scene with Jack Nicholson in the movie *Something's Gotta Give* that reminds us that patients shouldn't combine **sildenafil** with nitrates like **nitroglycerin** or they will end up in the emergency room.

The brand **Viagra** brings viable growth, taking the v-i-a from viable and –g-r from growth, – an erection.

2) Tadalafil (brand, Cialis)
ta-DAL-uh-fill (see-AL-is)

Again, the "afil, a-f-i-l" stem specifies the P-D-E-5 inhibitor class. **Cialis** is the weekend pill because it, unlike **sildenafil**, lasts the weekend with a long half-life.

I asked some students how they remembered **tadalafil**. I'm hesitant to share their mnemonic.

One said you just think "ta-dah" as in "surprise," to remember the first two syllables "tada, t-a-d-a."

I stopped them before they started on to how they remember the "fil" part of the generic name.

Cialis is the dual bathtub commercials drug.

Memorizing Pharmacology

Two Alpha Blockers for Benign Prostatic Hyperplasia (BPH)

1) Tamsulosin (brand, Flomax)
tam-syoo-LOW-sin (FLOW-Max)

Flomax allows for a flow of maximum urine taking the f-l-o from flow and m-a-x from maximum; **Flomax**.

The "osin" ending is not an actual stem, but a way to connect **tamsul*osin*** and **alfuz*osin*** as similar.

2) Alfuzosin (brand, Uroxatral)
al-fyoo-ZOH-sin (YUR-ox-uh-trall)

Again, the "osin, o-s-i-n" ending is not an actual stem, but a way to connect **tamsulosin** and **alfuzosin**.

With BPH, I think the brand **Uroxatral** sounds a little like "Urine control," taking the u-r from urine and "t-r-o-l, t-r-o-l" sound from control to match Uroxatral's t-r-a-l. **Uroxatral**.

A Pair of 5-Alpha-Reductase Inhibitors for Benign Prostatic Hyperplasia (B-P-H)

1) Duta*steride* (brand, Avodart)
due-TAS-ter-ide (AH-vo-dart)

Use the "–steride, s-t-e-r-i-d-e" stem to recognize the 5-alpha-reductase inhibitors like **dutasteride**. The "ster, s-t-e-r" for steroid helps you remember it's for men (and prostate).

2) Fina*steride* (brand names Proscar and Propecia)
fin-AS-ter-ide" (PRO-scar, pro-PEE-shuh)

Also employ the "–steride, s-t-e-r-i-d-e" stem to recognize the 5-alpha-reductase inhibitor **finasteride**

To connect the brand name **Proscar** to the generic **finasteride**, a professor told me of a student who used the phrase "That pro's car is the finest ride;" **Pros-car, finasteride**

140

The brand **Proscar** is for prostate care, taking the p-r-o from prostate and c-a-r from care. **Finasteride's** other brand name, **Propecia**, alludes to hair growth and is to reverse of alopecia (hair loss); pro-pecia (hair gain).

CHAPTER 7 QUIZ: ENDOCRINE, DRUG STEM AND DRUG CLASSIFICATION PRACTICE

I'm going to read a generic name for you and I want you to think about the stem, if any, and the class of medication.

1. Regular Insulin, no stem, short acting insulin
2. NPH Insulin, no stem, intermediate acting insulin
3. Levonorgestrel, gest, g-e-s-t, progestin used as emergency contraception
4. Metformin, formin, f-o-r-m-i-n, biguanide oral antidiabetic
5. Sitagliptin, gliptin, g-l-i-p-t-i-n, dipeptidyl peptidase-4, DPP-4, inhibitor
6. Glipizide, gli, g-l-i, 2nd generation sulfonylurea
7. Glyburide, gly, g-l-y, 2nd generation sulfonylurea
8. Glucagon, no stem, for hypoglycemia
9. Insulin lispro, no stem, rapid acting insulin
10. Insulin glargine, no stem, long acting insulin
11. Levothyroxine, no stem, but the thyro, t-h-y-r-o, in the middle should help you know it's for *hypo*thyroid
12. Propylthiouracil, uracil, u-r-a-c-i-l, for *hyper*thyroid
13. Testosterone, ster, s-t-e-r, steroid
14. Ethinyl estradiol with Norethindrone and Ferrous fumarate (brand name Loestrin 24 Fe), estr, e-s-t-r, estrogen, combined oral contraceptive with iron
15. Ethinyl estradiol with Norgestimate (brand name Tri-Sprintec), estr, e-s-t-r, estrogen with gest, g-e-s-t, progestin, combined oral contraceptive

16. Ethinyl Estradiol with Etonogestrel (brand name NuvaRing), estr, e-s-t-r, estrogen with gest, g-e-s-t, progestin, contraceptive vaginal ring
17. Ethinyl Estradiol with Norelgestromin (brand name OrthoEvra), estr, e-s-t-r, estrogen with gest, g-e-s-t, progestin, contraceptive patch
18. Oxybutynin, no stem, overactive bladder O-A-B
19. Solifenacin, fenacin, f-e-n-a-c-i-n, overactive bladder, O-A-B
20. Tolterodine, no stem, overactive bladder, O-A-B
21. Bethanechol, no stem, but note the "chol" c-h-o-l for cholinergic for urinary retention
22. Sildenafil, afil, -a-f-i-l, P-D-E-5 inhibitor
23. Tadalafil, afil, -a-f-i-l, P-D-E-5 inhibitor
24. Alfuzosin, no stem, alpha blocker for benign prostatic hyperplasia, B-P-H
25. Tamsulosin, no stem, alpha blocker for benign prostatic hyperplasia, B-P-H
26. Dutasteride, steride, s-t-e-r-i-d-e, 5-alpha-reductase inhibitor for BPH
27. Finasteride, steride, s-t-e-r-i-d-e, 5-alpha-reductase inhibitor for BPH and hair loss (alopecia)

ENDOCRINE AND MISCELLANEOUS: MEMORIZING THE CHAPTER

"E" ENDOCRINE

Start with the OTC two middle peak insulins from the "r" and "n" in endoc<u>r</u>i<u>n</u>e: **regular insulin** and **insulin NPH**; then Plan B One-Step **levonorgestrel**. RX: four oral antidiabetics in order of drug class: the biguanide **metformin,** the DPP-4 **sitagliptin,** and two sulfonylureas **glipizide** and **glyburide**; then **glucagon** for when the <u>gl</u>ucose is <u>go</u>ne. On to prescription shortest- and longest- acting insulins: **insulin lispro (brand name, Humalog)** and **insulin glargine (brand**

CHAPTER 7: ENDOCRINE and MISCELLANEOUS

name, **Lantus**). "T" in Lan*tus* to "T" for thyroid level low (**levothyroxine**), to thyroid level high **(PTU).** "T" again for low T, **testosterone**. Testosterone "-ster-" (s-t-e-r) stem to estrogen "estr-" (e-s-t-r) stem. **Ethinyl estradiol** "estr-" times four in four oral contraceptives, from high on the body to low, starting with an oral (po) form with iron, another p.o. tri-phasic, a belly patch, and vaginal ring - brand names are easier first, Loestrin 24 Fe, Tri-Sprintec, Ortho-Evra patch, NuvaRing. Generic progestins are next: **norethindrone, norgestimate, norelgestromin, etonogestrel**. Overactive bladder (OAB) **oxybutynin, solifenacin,** and **tolterodine**, to the opposite – urinary retention: **bethanechol**; to a lack of erection by half-life: **sildenafil** short and **tadalafil** long, but also **tadalafil** for BPH and alpha-blockers **alfuzosin** and **tamsulosin**, and 5-alpha reductase **dutasteride** and **finasteride**.

Memorizing Pharmacology

Can you recite these in order?:

Regular Insulin	Ethinyl Estradiol /	Oxybutynin
NPH Insulin	Norethindrone /	Solifenacin
	Ferrous fumarate	Tolterodine
Levonorgestrel	(Loestrin 24 Fe)	
		Bethanechol
Metformin	Ethinyl Estradiol /	
Sitagliptin	Norgestimate	Sildenafil
Glipizide	(Tri-Sprintec)	Tadalafil
Glyburide		
	Ethinyl Estradiol /	Alfuzosin
Glucagon	Etonogestrel	Tamsulosin
	(NuvaRing)	
Insulin lispro		Dutasteride
Insulin glargine	Ethinyl Estradiol /	Finasteride
	Norelgestromin	
Levothyroxine	(OrthoEvra)	
Propylthiouracil		
Testosterone		

Once you're able to recite the list, you are ready to move on.

CHAPTER 8: MEMORIZING THE BOOK

MEMORIZING THE BOOK PART I

Here's a complete version of the entire 200 starting with the seven "Memorizing the Chapters." If you just want to hear the 200 drug names, you can skip to the next track.

"G" GASTROINTESTINAL

Broadly speaking, there are 13 drugs in the gastrointestinal section, six pairs and a single agent.

Picture in your mind where in the human body these work. Start with the six acid reducers in the stomach, move down to the where the two laxatives and antidiarrheals work in the intestinal tract, and go back up to the antiemetics to prevent vomiting from the mouth, and back down to the colon to the ulcerative colitis medication.

Why this up and down and up and down? It's easiest to start at the stomach and work down to the intestines with O-T-C drugs, then work top to bottom from the mouth to the intestines with the prescription only drugs.

Remember the antacids works the quickest followed by the H2 blockers and the proton pump inhibitors and we'll memorize them in that order.

The antacids **calcium carbonate** and **magnesium hydroxide** are both in the same column on the periodic table of elements in alphabetical order. Two H_2 Blockers: **famotidine** and **ranitidine** follow. Two PPIs **esomeprazole** and **omeprazole** follow them. Often diarrhea follows an upset stomach, so we treat with **bismuth sub<u>sa</u>licylate** and **loperamide**. Use "L" from loperamide to get to "L" for two laxatives: **docusate sodium** and **polyethylene glycol**. Use the "p-o" from polyethylene reversed as "o-p" for **on<u>dans</u>etron** and **promethazine**. You can rectally administer promethazine next to the colon to get to the ulcerative colitis medication – **inf<u>li</u>ximab**.

"M" Musculoskeletal

Three OTC NSAIDs, **aspirin, ibu<u>prof</u>en, naproxen,** and then **acetaminophen,** a non-narcotic analgesic. Combine **aspirin with acetaminophen** and **caffeine** to make **Excedrin Migraine.** Use migraine (M) and caffeine (C) to connect **melox<u>ic</u>am** (M) and **cele<u>coxib</u>** (C). Group opioids by DEA class. Start with DEA Schedule II **morphine**, the original agent; then **fentanyl, hydrocodone with acetaminophen** and **oxycodone with acetaminophen.** Next comes DEA schedule III, **acetaminophen with codeine**; to DEA Schedule IV **tram<u>adol</u>**; then to the opioid *narcotic antagonist* **na<u>lox</u>one (brand name, Narcan);** to migraine agony *agonists* for migraine headache with **ele<u>trip</u>tan** and **suma<u>trip</u>tan;** down from the head to rheumatoid arthritis inside the joints with the DMARD **methotrexate**, a non-biologic; to two biologic DMARDs; **aba<u>tac</u>ept** to **eta<u>ner</u>cept**. From joint pain on to the brittle bones with **alen<u>dron</u>ate** and **iban<u>dron</u>ate** for osteoporosis. Then move out to the muscles with the muscle relaxers **cyclobenzaprine** and **di<u>aze</u>pam** all the way down to the big toe for the gout meds **allopurinol** and **febu<u>xos</u>tat**.

"R" Respiratory

Start with the 1st-generation O-T-C antihistamine **diphenhydramine**, and then go to the 2nd-generation **cetirizine** (which starts with a C) then alphabetically to 2nd-generation **lora<u>ta</u>dine** (with an L). Add a decongestant to make **lora<u>ta</u>dine-D** behind-the-counter (B-T-C), and then move to what the "hyphen D" stands for – **pseudoephedrine,** the decongestant alone. Then walk into the OTC aisle to get the oral or nasal decongestant **phenylephrine** and move up to the nose with the intranasal-only decongestant **oxymetazoline**. Stay in the nose with an OTC intranasal glucocorticoid **triamcinolone**. Move from *nasal* congestion to *chest* congestion with **guaifenesin with dextromethorphan** to another antitussive combination behind the prescription counter for prescription-only, **guaifenesin with codeine**. If that doesn't work and the coughing inflames your lungs, you might need an oral steroid like **methylprednisolone** (which starts with an M and has "pred" p-r-e-d in the middle) or **prednisone** (which follows alphabetically, beginning with P). After the acute attack, you might find you have asthma and need to use a prophylactic inhaled steroid in combination with a long-acting beta$_2$ receptor agonist, either **budesonide** with **formoterol** or **salmeterol** with **fluticasone**. You could individually use the steroid **fluticasone** or **albuterol,** the short-acting beta 2 receptor agonist. The beta 2 receptor agonist **albuterol** combined with anticholinergic **ipratropium** makes the duo in the brand, **DuoNeb**, or alternatively, the long-acting anticholinergic **tiotropium** can be given alone. If that doesn't work, use **montelukast** against leukotrienes or **omalizumab** against IgE, but remember that **omalizumab** has a black box warning about anaphylaxis, which might necessitate an injection of **epinephrine**.

"I" IMMUNE

The sub-algorithm is to start with antibacterial agents, then go to antifungals, then antivirals, which are in alphabetical order

b for bacterial
f for fungal
and v for viral

of what they are "anti-;" anti-bacterials, anti-fungals, anti-virals.

So first the OTC antibacterial cream **neomycin** with **polymyxin-B and bacitracin**; **butenafine**, the antifungal cream; and the prophylactic antiviral **influenza vaccine**; followed by an acute antiviral – **docosanol**.

Begin again with systemic antibacterials, first those that attack the cell wall, the beta lactamase susceptible amino penicillin **amoxicillin**, followed by the augmented beta lactamase resistant **amoxicillin** with **clavulanate.**

Move to cephalosporins in generational order: 1st generation **cephalexin**; to 3rd generation **ceftriaxone**;

to 4th generation **cefepime**, the "maximum" generation; to alphabetically last "v" for the glycopeptide antibiotic **vancomycin** for MRSA.

From bactericidal cell wall inhibitors to bacteriostatic inhibitors of protein synthesis, the tetracyclines: **doxycycline** and **minocycline** followed by three macrolides in order of the number of times taken per day: qd (once daily), b-i-d, (twice daily), q-i-d, (four times daily), **azithromycin, clarithromycin**, and **erythromycin** respectively.

Use the "-mycin" ending to get to **clindamycin** and the "l-i-n" from c*lin*damycin to go to *linezolid*.

Follow with the bactericidal inhibitors of protein synthesis, the aminoglycosides **amikacin** and **gentamicin**.

Move to the UTI medications **sulfamethoxazole** and **trimethoprim, ciprofloxacin**, and **levofloxacin;** and from "l" in "levo" to "m" in **metronidazole**, the antiprotozoal.

Then four letters for four TB drugs because over 4 mm induration is a positive TB test or a ripe, "r-i-p-e" result: (R) in **rifampin**, (I) in **isoniazid**, (P) in **pyrazinamide**, and (E) in **ethambutol**.

TB patients often have opportunistic fungi, so three antifungals alphabetically follow: **amphotericin B**, **flu<u>co</u>nazole**, and **nystatin.**

The antivirals for influenza follow those: **oselt<u>ami</u>vir** and **zan<u>ami</u>vir**; then **acy<u>clo</u>vir** and **valacy<u>clo</u>vir** for HSV and VZV, in order of half-life; then for RSV **pal<u>ivi</u>zumab**; then for HIV, in order of attack: fusion, CCR5, reverse transcriptase, integrase, and protease. I have to use brand names to memorize these first because those are easier: **Fuzeon, Selzentry, Atripla, Isentress, Prezista**, and then I attach the generic names for those: **enfu<u>vir</u>tide, mara<u>vir</u>oc**, (**efa<u>vir</u>enz** with **emtri<u>cita</u>bine** and **tenofo<u>vir</u>**), **ralte<u>gra</u>vir** and **daru<u>nav</u>ir.**

"N" Neuro

The four neuro OTC drugs are the local anesthetics **benzo<u>caine</u>** and **lido<u>caine</u>**, which are an ester and amide respectively. **Meclizine** is for dizziness, and **acetaminophen** with **diphenhydramine** is for insomnia. Connect this with other anti-insomniacs the benzodiazepine-like sedative-hypnotics **eszopi<u>clone</u>, zolpidem**, and melatonin receptor agonist **ramelteon**. No sleep equals depression. Five SSRI antidepressants: **citalopram** and **escitalopram, sertraline** in the middle followed by three "–oxetines:" **flu<u>oxetine</u>** and **par<u>oxetine</u>** (the SSRIs) and **dul<u>oxetine</u>** (the SNRI) and **venlafaxine** the SNRI; followed by, in order of safety, the TCA **ami<u>trip</u>tyline** and the M-A-O-I **isocarboxazid**. Move from depression to smoking: **bupropion** and **vare<u>nic</u>line,** smoking while anxious, two –azolams: **alp<u>ra</u>zolam** and **mid<u>azo</u>lam** and two -azepams **clon<u>aze</u>pam** and **lor<u>aze</u>pam**. From the "A" in anxious to the "A" in ADHD stimulants **dexmethylphenidate, methylphenidate**, to non-stimulant **atom<u>ox</u>etine**; to "m-o" from atomoxetine for m-o-o-d, mood stabilizer **lithium**; then "L" from lithium to <u>l</u>ow potency 1st-generation **chlorpromazine**, to high potency "halo" **halo<u>peridol</u>,** "–peridol" to "–peridone," 2nd generation whisper **risperidone** to whisper-quiet **quetiapine**; pine to traditional antiepileptics **carbamazepine, divalproex** and **phenytoin** to two newer antiepileptics Neu-rontin (which is **gabapentin**) and **pregabalin**. From epileptic motion to Parkinsonian motion: **carbidopa with levodopa** and **selegiline** senility to Alzheimer's "memory is done" taking the m-e-m from **memantine** to d-o-n-e **donepezil** ending with "D" dizzy **scopolamine**.

"C" Cardio

"O" from cardio to **Omega-3 acid ethyl esters** to **niacin** to **aspirin** you need to take before the niacin to prevent flushing. Flushing five diuretics in nephron order: **mannitol, furosemide, hydrochlorothiazide, hydrochlorothiazide with triamterene, spironolactone** (potassium sparing), then **potassium chloride** to alphas: alpha 1 **doxazosin**, alpha 2 **clonidine**; to betas: 1st-generation beta 1 and beta 2 non-selective **propranolol**; 2nd-generation, beta 1 only, **atenolol, metoprolol** tartrate short acting to **metoprolol** succinate long acting; to 3rd-generation **carvedilol**. Dil to pril ACE inhibitors **enalapril, lisinopril**. ARBs with letters "love," "l-o-v;" **losartan, olmesartan** and **valsartan**; from "-sartan" to CCBs non-dihydropyridines **diltiazem** and **verapamil** to the dihydropyridines **amlodipine** and **nifedipine** vasodilating only; CCBs to **nitroglycerin**, a vasodilator. Nitroglycerin brand name **Nitrostat** to stat-ins: **atorvastatin** to **rosuvastatin**; LDL to lowered triglycerides, **fenofibrate** -- fibs children tell to parents, parenteral **enoxaparin** and **heparin**, "-parin" to "-farin," **warfarin** enteral anticoagulant with **dabigatran**, "d" to **digoxin** for CHF and **atropine** to prevent a bradycardic mess.

"E" Endocrine

Start with the OTC two middle peak insulins from the "r" and "n" in endocrine: **regular insulin** and **insulin NPH**;

then Plan B One-Step **levonorgestrel**. RX: four oral antidiabetics in order of drug class: the biguanide **metformin,** the DPP-4 **sitagliptin,** and two sulfonylureas **glipizide** and **glyburide**; then **glucagon** for when the glucose is gone. On to prescription shortest- and longest- acting insulins: **insulin lispro (brand name, Humalog)** and **insulin glargine (brand name, Lantus).** "T" in Lan*tus* to "T" for thyroid level low **(levothyroxine)**, to thyroid level high **(PTU).** "T" again for low T, **testosterone**. Testosterone "-ster-" (s-t-e-r) stem to estrogen "estr-" (e-s-t-r) stem. **Ethinyl estradiol** "estr-" times four in four oral contraceptives, from high on the body to low, starting with an oral (po) form with iron, another p.o. tri-phasic, a belly patch, and vaginal ring -- brand names are easier first, Loestrin 24 Fe, Tri-Sprintec, Ortho-Evra patch, NuvaRing. Generic progestins are next: **norethindrone, norgestimate, norelgestromin, etonogestrel**. Overactive bladder (OAB) **oxybutynin, solifenacin,** and **tolterodine**, to the opposite – urinary retention: **bethanechol**; to a lack of erection by half-life: **sildenafil** short and **tadalafil** long, but also **tadalafil** for BPH and alpha-blockers **alfuzosin** and **tamsulosin**, and 5-alpha reductase **dutasteride** and **finasteride**.

Memorizing the book Part II

And here are the 200 drugs without the mnemonics.

Calcium carbonate	Bismuth sub<u>sali</u>cylate	In<u>fli</u>ximab
Magnesium hydroxide	Loperamide	
	Docusate sodium	
Famo<u>tidine</u>	Polyethylene glycol	
Rani<u>tidine</u>		
	Ondan<u>setron</u>	
Esome<u>prazole</u>	Promethazine	
Omeprazole		

Aspirin	Morphine	Metho<u>trexate</u>
Ibu<u>profen</u>	Fentanyl	Aba<u>tacept</u>
Naproxen		Eta<u>nercept</u>
	Hydrocodone with	
Acetaminophen	acetaminophen	Alen<u>dronate</u>
	Oxycodone with	Iban<u>dronate</u>
Aspirin with	acetaminophen	
Acetaminophen		Cyclobenzaprine
And Caffeine	Acetaminophen	Di<u>azepam</u>
	with codeine	
Melox<u>icam</u>		Allopurinol
	Tram<u>adol</u>	Febu<u>xostat</u>
Cele<u>coxib</u>		
	Na<u>l</u>oxone	
	Ele<u>triptan</u>	
	Suma<u>triptan</u>	

Diphenhydra-mine	Triamcinolone	Albu<u>terol</u> <u>alone</u>
Cetirizine	**Guaifenesin with dextromethorphan**	Albu<u>terol</u> with Ipra<u>tropium</u>
Lor<u>atadine</u>	Guaifenesin with codeine	Tio<u>tropium</u>
Lor<u>atadine</u>-D Pseudoephed-rine	Methyl<u>pred</u>nisolone <u>Pred</u>nisone	Monte<u>lukast</u> Oma<u>lizumab</u>
Phenylephrine Oxymetazoline	Budesonide with Form<u>oterol</u> Fluticasone with Salm<u>eterol</u> Fluticasone alone	Epinephrine

CHAPTER 8: MEMORIZING THE BOOK

Neomycin with Polymyxin-B and Bacitracin	Azi<u>thromycin</u> Clari<u>thromycin</u> Ery<u>thromycin</u>	Amphotericin B Flu<u>con</u>azole Nystatin
Butenafine	Clind<u>amycin</u> Line<u>zolid</u>	Oselt<u>amivir</u> Zan<u>amivir</u>
Influenza vaccine		
Docosanol	Am<u>ikacin</u> Genta<u>micin</u>	A<u>cyclovir</u> Vala<u>cyclovir</u>
Amoxi<u>cillin</u> Amoxi<u>cillin</u> with Clavulanate	<u>Sulfa</u>methoxazole with Trimetho<u>prim</u>	Pali<u>vi</u>zumab Enfu<u>vir</u>tide Mara<u>vir</u>oc
<u>Cephalexin</u> <u>Cef</u>triaxone <u>Cef</u>epime	Cipro<u>floxacin</u> Levo<u>floxacin</u>	Efa<u>vir</u>enz with Emtri<u>cit</u>abine and
Vanco<u>mycin</u>	Metro<u>nidazole</u>	Tenofo<u>vir</u> Daru<u>nav</u>ir
Doxy<u>cycline</u> Mino<u>cycline</u>	<u>R</u>ifampin Isoniazid Pyrazinamide Ethambutol	Raltegravir

Benzocaine	Ami<u>triptyline</u>	Chlorpromazine
Lidocaine		Halo<u>peridol</u>
	Isocarboxazid	
Meclizine		Ris<u>peridone</u>
	Bupropion	Que<u>tiapine</u>
Acetaminophen PM	Vare<u>nicline</u>	
		Carbamaze<u>pine</u>
Eszopi<u>clone</u>	Alpr<u>azolam</u>	Divalproex
Zol<u>pidem</u>	Mid<u>azolam</u>	Pheny<u>toin</u>
Ra<u>melteon</u>		
	Clon<u>azepam</u>	G<u>a</u>bapentin
Citalopram	Lor<u>azepam</u>	Preg<u>a</u>balin
Escitalopram		
Ser<u>traline</u>	Dexmethylpheni-	Levo<u>dopa</u> with
Flu<u>oxetine</u>	date	carbi<u>dopa</u>
Par<u>oxetine</u>	Methylphenidate	Sele<u>giline</u>
	Atom<u>oxetine</u>	Memantine
Dul<u>oxetine</u>		Done<u>pezil</u>
Venla<u>faxine</u>	Lithium	
		Scopolamine

Omega-3-Acid Ethyl Esters	Propran<u>olol</u>	<u>Nitro</u>glycerin
Niacin	Aten<u>olol</u>	
	Metop<u>rolol</u> Succinate	Ator<u>vastatin</u>
	Metop<u>rolol</u> Tartrate	Rosu<u>vastatin</u>
Aspirin (Low Dose)	Carve<u>dilol</u>	
		Feno<u>fibrate</u>
Mannitol	Enala<u>pril</u>	
Furo<u>semide</u>	Lisino<u>pril</u>	He<u>parin</u>
Hydrochloro<u>thiazide</u>		Enoxa<u>parin</u>
Hydrochlorothiazide with triamterene	Lo<u>sartan</u>	
Spironolactone	Olme<u>sartan</u>	War<u>farin</u>
	Val<u>sartan</u>	Dabi<u>gatran</u>
Potassium Chloride	Dil<u>tia</u>zem	Clopido<u>grel</u>
	Vera<u>pamil</u>	
Dox<u>azosin</u>		Digoxin
Clonidine	Amlo<u>dipine</u>	
	Nife<u>dipine</u>	Atropine

Regular Insulin	Ethinyl Es<u>tr</u>adiol /	Oxybutynin
NPH Insulin	Norethindrone /	Soli<u>fen</u>acin
	Ferrous fumarate	Tolterodine
Levonor<u>ges</u>trel	(Loestrin 24 Fe)	
		Bethanechol
Met<u>form</u>in	Ethinyl Es<u>tr</u>adiol /	
Sita<u>glipt</u>in	Nor<u>ges</u>timate	Silden<u>afil</u>
<u>Gl</u>ipizide	(Tri-Sprintec)	Tadal<u>afil</u>
Gl<u>y</u>buride		
	Ethinyl Es<u>tr</u>adiol /	Alfuzosin
Glucagon	Etono<u>ges</u>trel	Tamsulosin
	(NuvaRing)	
Insulin lispro		Duta<u>steride</u>
Insulin glargine	Ethinyl Es<u>tr</u>adiol /	Fina<u>steride</u>
	Norel<u>ges</u>tromin	
Levothyroxine	(OrthoEvra)	
Propylthiouracil		
Testo<u>ster</u>one		

CHAPTER 9: END OF SEMESTER

At the end of term, there is an awkward and hurried separation. Instead of saying goodbye, or having some closing ritual, a student takes a test, leaves, and may never talk to the instructor again. Nevertheless, they have just taken their final exam.

Shouldn't we talk about it and acknowledge what they've learned over the term? These questions that follow are final exam questions from one of my old essay tests, much of the material comes from this book, but some is new.

No pressure here – that doesn't fit the tone of this book. I just want you to hear the question and see if you can answer it based on what you've learned in the book, but most importantly, does the answer make sense to you now that you know all of these drugs' names? This book ultimately is about making you fluent in the language of pharmacology. The following questions will challenge your fluency.

INTRODUCTION - BASIC PRINCIPLES

1. Provide definitions for affinity, intrinsic activity, agonist and antagonist.

Affinity – while affinity in lay language comes from the Latin "ad" and "finis," meaning "to a border" it later came to signify a relationship. In pharmacology and more generally, biochemistry, it indicates a drug's attraction to a receptor.

Intrinsic Activity – how well a drug produces the greatest response. An analogy for the difference between affinity and intrinsic activity would be the relationship between two people. A person might have an affinity for someone else, like him or her, but produce no intrinsic activity. The other person doesn't like them back.

Agonist – a chemical that binds to a receptor to produce a response. Be careful, this does not mean it's *good* or *bad*; simply that it activates a receptor. For example, a person can turn a light on or off. Light is good for visibility, darkness is good when developing photographs. The hand (the agonist) activates the receptor (light switch) in both cases.

Antagonist – a chemical that binds to a receptor to block the agonist. For example, you wanted to take someone to prom, but that person decided to take her brother because she doesn't want to deal with any drama. The brother serves as the antagonist; he gets in the way. Other analogies include:

Protestors blocking your way to work or school.
A bathtub drain plug preventing water from draining.
A cornerback blocking a catch to a wide receiver.

2. Match the four pharmacokinetic principles with their appropriate organ systems (one is actually a tissue, not an organ).

First, you need to understand the four pharmacokinetic principles: **absorption, distribution, metabolism, and excretion** (sometimes referred to as AD-ME, A-D-M-E). Imagine traveling through the body's gastrointestinal tract.

A. Absorption – First, if you take the medication orally, it goes through the stomach and then into the small intestine. It's a mistake to think that the stomach is the area of greatest absorption. Villi, hair-like projections, line the *intestines* and greatly increase the surface area for drug absorption.

D. Distribution – *Blood* is a tissue and transports medication throughout the body. The circulatory system works as the paperboy and delivers newspapers along a route.

M. Metabolism – Think of the words *anabolism* (to build up, like anabolic steroids) and *catabolism* (break down), then see how the *meta* fits in. In Greek, meta means "after" or "beyond." The *liver* often produces these changes; therefore, *metabolism* explains the product "after" the liver.

E. Excretion – While we have many sites of excretion (sweat, breast milk, feces, and kidneys), generally drug removal and excretion occurs through the *kidneys* in urine.

3. Provide four medications, two with matching stems in the prefix position, two with matching stems in suffix position. What does having a matching (United States Adopted Names Council Approved) stem mean about the relationship between the drugs?

Cefepime and **ceftriaxone** both have "cef-"at the beginning of the generic medication names. Two drugs with similar names may or may not be related. To be sure, a person would have to consult the list put out by the United States Adopted Names Council currently found on the American Medical Association's (A-M-A) website. In this case, "cef, c-e-f"is an official stem for "cephalosporins."

However, to say that both **famotidine** and **loratadine** have the same endings does not mean the same thing as two drugs having "cef-"as prefixes. While both end in "ine, i-n-e" they are not in the same class of medications.

The USANC recognizes **"tidine, t-i-d-i-n-e"** as the stem for H_2-receptor antagonists, cimetidine type, a medication for gastrointestinal issues and recognizes **"atadine, a-t-a-d-i-n-e"** as the *tricyclic histaminic-H1 receptor antagonists, loratadine derivatives with the former stem "tadine, t-a-d-i-n-e"*.

In plain English, that means if you connect these two drugs by only "ine, i-n-e" you assert an H_2 blocker for acid does the same thing an H_1 blocker does for allergies, which is not true.

A correct example of medications with two USANC approved stems in the suffix positions includes **pen*icillin*** and

Memorizing Pharmacology

amoxi*cillin*, both penicillin class antibiotics with a "cillin, c-i-l-l-i-n" stem.

4. What does metabolic inhibition and metabolic induction do to the level of a drug in the body?

Metabolic induction increases metabolism and decreases drug levels while metabolic inhibition slows metabolism and increases drug levels. When you think of inhibiting something, it means to stop it from happening. An A-C-E-I like **enalapril** is an angiotensin converting enzyme inhibitor. Therefore, this medication stops the angiotensin-converting enzyme from working. In the same way, metabolic inhibition is to stop or inhibit metabolism. This is a little tricky because there is an inverse relationship. If you increase inhibition, you stop the factory, usually in the liver, from metabolizing a drug – breaking it down or changing it. Then, you will have *more* drug around because *less* is broken down.

Induction is a word you've probably heard with labor. To "induce labor" is to make it happen faster. That can be a good thing. If a woman goes too far beyond the normal human 40-week term, there can be serious consequences for the fetus. In this case, induction is to make the factory, usually in the liver, break down or change, metabolize, medication faster. If you break down medications faster, you have *less* drug in the circulation.

5. Provide the names of two pairs of drugs with the same root drug, not stem in the prefix or suffix position, but one is the S- isomer of the other. How does the S- affect the efficacy of the medication and what is its Latin origin?

The same root in a medication means that it's probably racemic. This is a term from organic chemistry. For example, **esomeprazole** and **omeprazole** both have the same root as do **escitalopram** and **citalopram**. The S- and R- isomers of a chemical or drug are mirror images of each other and sometimes one is effective and one is not. In the case of these four

CHAPTER 9: END OF SEMESTER

medications, the S- isomer is the effective one. **Omeprazole** and **citalopram** are both combinations of S- and R-, meaning they contain the ineffectual R- isomer. We use the letters "e-s" instead of "s" because by only using the letter "s," **citalopram** would become "scitalopram, s-c-i-t-a-l-o-p-r-a-m" giving two different drugs identical pronunciation. The "s" comes from the Latin "sinister," which means left, because people who were left-handed were thought to be sinister. The "r" means "rectus," which is Latin for right.

CHAPTER 1 - GASTROINTESTINAL

1. Provide an example of an antacid, H2 blocker, and proton pump inhibitor and match a unique disease state or condition to each medication.

Antacids include medications such as **magnesium hydroxide** or **calcium carbonate**. Patients use both for hyperacidic states, but **magnesium hydroxide** relieves constipation and **calcium carbonate** supplements calcium deficiency. Patients use H$_2$-blockers such as **famotidine** or **ranitidine** for conditions such as gastroesophageal reflux disease (GERD, G-E-R-D), with the advantage of less frequent dosing than antacids. A triple drug regimen includes a PPI such as **esomeprazole** for peptic ulcer disease (P-U-D).

2. Contrast one over-the-counter anti-diarrheal and one prescription antidiarrheal and the reason for choosing one over the other.

Patients with diarrhea, but without an active GI infection, use over-the-counter antidiarrheals like **loperamide** or **bismuth subsalicylate**.

Prescribers employ antibiotics and rehydration to treat infectious diarrhea. Antidiarrheals keep toxins in the patient's gastrointestinal system.

3. Contrast an over-the-counter anti-nausea and one prescription only anti-nausea mediation and provide a specific appropriate use for each one.

Patients use OTC **meclizine** for motion sickness. Its old brand name **Antivert** implies anti-vertigo. A prescription medication like **ondansetron** prevents chemotherapy-induced nausea and vomiting (C-I-N-V). Patients take this prior to chemotherapy.

4. Define chelation. Provide two medications from separate drug classes that would chelate together and describe the impact of this on the drugs' effectiveness.

The binding of one chemical to another, involving a metal, is chelation. We might not think of **calcium** and **magnesium** as metals, especially when we find them in a **Tums** tablet or **Milk of Magnesia** liquid, but they bind to some medications, especially antibiotics like **doxycycline** and **ciprofloxacin**.

A good way to think of this is to use the example of table salt, sodium chloride (N-a-C-l). When isolated, sodium (N-a) is caustic and explosive. We use chlorine (C-l) in pools and it can be very toxic as well. However, put sodium and chloride together and they form an inert entity, suitable for popcorn and consumption.

We don't want an inert antibiotic, but with chelation of **doxycycline** with magnesium, calcium, or aluminum, that's what we might get. As such, prescribers recommend patients limit dairy and antacids to at least a half-hour before or two hours after ingestion of the medication.

5. Provide two drug examples of laxative medications.

The osmotic agent **polyethylene glycol** or stool softener **docusate** combats constipation often caused by opioid analgesic use.

Chapter 2 - Musculoskeletal

1. Compare the potency and analgesic efficacy of prescription strength ibuprofen 800 mg and naproxen 500 mg.

Prescription **ibuprofen** 800mg is about as potent as 500 mg of **naproxen**. Sometimes a patient will look to get the "strongest" medicine and mistake the higher milligram strength for a stronger medicine. Two medications that are equipotent, equal in potency, have no advantage in analgesic efficacy, how effective a pain medicine is, based on milligram strength.

Since it's a tie for efficacy, a practitioner might look to the daily dosing requirements to make a choice. A patient would need to take **ibuprofen** four times a day whereas we dose **naproxen** twice daily. It's easier to remember to take two tablets daily. This is **naproxen's** advantage, not its potency.

2. Identify three medications, one for each level of the pain scale at ten, five and three. Provide a rationale for their particular use and the DEA class of each medication, if any.

While the pain scale uses one-point increments from zero (no pain) to ten (worst possible pain), it's convenient to divide it into three general regions: about ten, five, and three.

Opioid analgesics like morphine relieve pain at the highest level. **Morphine** is DEA Schedule II, the most addicting class of legal meds. **Hydrocodone** with **acetaminophen** falls around the middle of the pain scale. It's also DEA Schedule II. OTC **Ibuprofen** provides relief at the bottom of the scale.

3. Contrast a DMARD and an NSAID. Provide an example of each, and rationale for their use.

A DMARD stands for Disease Modifying Anti-Rheumatic Drug. The DMARD **methotrexate** suppresses the immune response in an autoimmune disease like rheumatoid arthritis. An NSAID is a non-steroidal anti-inflammatory drug. The NSAID **ibuprofen** provides relief from inflammation or osteoarthritis.

4. Rationalize one therapeutic use of A-S-A over A-PAP and A-PAP over A-S-A.

A-S-A stands for acetylsalicylic acid, or **aspirin**. **Aspirin** relieves inflammation, thins platelets, provides analgesia, and reduces fever. **Acetaminophen** provides analgesia and reduces fever, but does not affect platelets nor provide an anti-inflammatory response. Pregnant patients might use **acetaminophen** as **aspirin** poses significant risk to the fetus. Children with fever use **acetaminophen** rather than **aspirin** to avoid Reye's syndrome or an increased bleeding risk.

5. Provide a rationale for each of the three medications found in Excedrin Migraine.

Excedrin contains **aspirin**, **acetaminophen**, and **caffeine** and each ingredient relieves the pain of a headache. **Aspirin** helps curb inflammation and acts as an analgesic. A drug like **acetaminophen** helps with analgesia, but without additive adverse gastrointestinal effects. **Caffeine** causes vascular vasoconstriction, to reduce the pain of a swollen blood vessel in the brain. If headaches are the result of the inflammation and vasodilation in the brain, where there is no room to expand, then reducing inflammation and vasodilation would reverse these effects.

Chapter 3 - Respiratory

1. Provide an example of an over-the-counter and a prescription-only cough suppressant. Then describe a specific concern regarding the type of patient who gets the prescription-only cough suppressant.

An over-the-counter mucolytic with cough suppressant such as **guaifenesin with dextromethorphan** relieves a dry unproductive cough. If the cough doesn't respond to the over-the-counter remedy, we prescribe a liquid antitussive such as **guaifenesin** with **codeine**. Prescribers must use caution when providing this addictive DEA scheduled medication to patients with a history of narcotic abuse.

2. Contrast an H1 antihistamine with an H2 antihistamine, identifying an example of each and relate the therapeutic function of each to treatment of two specific pathophysiologic conditions.

An H_1 antihistamine clears seasonal allergy symptoms. The first-generation **diphenhydramine** sedates, whereas second-generation forms **loratadine** or **cetirizine** do not. We refer to anti-allergy medications as simply "antihistamines."

An H_2 antihistamine reduces acidity in patients with gastroesophageal reflux disease (GERD). Medications such as **famotidine** or **ranitidine** fall into this class. We refer to them as H_2-blockers to avoid confusion. Stay vigilant as the two stems, "atadine, a-t-a-d-i-n-e" and "tidine, t-i-d-i-n-e" differ by only two letters.

3. Give an example of a 1st- and a 2nd-generation antihistamine (generic and brand name) and provide an explanation for the difference between the adverse effects of the two.

A first-generation antihistamine such as **diphenhydramine** causes sedation. A second-generation antihistamine, such as

Memorizing Pharmacology

loratadine or **cetirizine**, does not. Second-generation antihistamines don't cross the blood-brain-barrier to introduce CNS suppression associated with drowsiness.

4. Identify the two major components of asthma and provide an example of a medication that would treat each component and note any specific roots / suffixes.

Asthma is a condition of bronchoconstriction and inflammation. As such, a medication that produced bronchodilation and an anti-inflammatory effect would be effective. **Fluticasone with salmeterol** (brand, **Advair**) manages asthma prophylactically. **Fluticasone** is a steroid that provides an anti-inflammatory effect.

While "sone, s-o-n-e" appears in many steroids, there are medications that have "sone" and are not steroids. For example, we classify **dapsone** as an antibiotic used for leprosy or pneumocystis pneumonia.

Salmeterol is a long-acting beta$_2$ agonist that provides bronchodilation. Using **fluticasone** with **salmeterol** does not preclude the need for **albuterol,** a rescue inhaler. Remember, the suffix "terol, t-e-r-o-l" indicates a beta$_2$ agonist, but does not identify short versus long acting medications in the class.

5. Provide three different steroidal medications used to manage asthma.

A severe asthma attack often requires **prednisone** to reduce the inflammation. Severe bronchial inflammation responds to **methylprednisolone**, a derivative of **prednisolone**. The **fluticasone** in brand **Advair** or **budesonide** in brand **Symbicort** reduces asthmatic severity and incidence.

Chapter 4 - Immune

1. Identify three unique macrolide medications, their dosing schedules and effect on compliance.

Three macrolides include **erythromycin** dosed four times daily, **clarithromycin**, dosed twice a day, and **azithromycin** dosed once daily with a doubled loading dose. Patient compliance improves with fewer doses.

2. Identify five major drug classes related to the common cold. Provide their therapeutic effects.

While a drugstore aisle of cold medications contains an overwhelming number of medications, five major drug classes comprise the bulk of the ingredients.

The *analgesic*, ibuprofen relieves fever, pain, and inflammation whereas acetaminophen only relieves fever and pain.

Antihistamines help with runny nose, watery eyes, sneezing. Sometimes **diphenhydramine** "P-M" provides a means to improve sleep.

The *decongestant* **pseudoephedrine** clears congestion. Patients often mistake decongestants for antihistamines. A decongestant doesn't help with allergy symptoms.

Antitussives such as **dextromethorphan**, e.g., the D-M in Robitussin, suppress cough.

The *mucolytic* **guaifenesin** found in **Robitussin** and **Mucinex** breaks up mucous. Mucolytic means "mucous" + "break down."

3. Identify three disadvantages of a first-generation cephalosporin versus a third-generation cephalosporin and the general mechanism of action of the cephalosporin class.

A first-generation cephalosporin such as **cephalexin** has poor penetration to the cerebrospinal fluid (C-S-F), weak gram-negative coverage, and poor resistance to beta-lactamases, specifically cephalosporinases.

A third-generation medication like **ceftriaxone** has the opposite characteristics, with strong gram-negative coverage, the ability to penetrate the C-S-F, and good action against beta-lactamases.

The general mechanism of the cephalosporin class resembles penicillin. They inhibit the formation of bacterial cell wall, cause the cell wall to leak and cause death. Think of a bubble popping. Thus, cephalosporins are bactericidal.

4. Identify the advantage of amoxicillin with clavulanate (brand, Augmentin) vs. amoxicillin alone. Explain the general mechanism of action of the penicillin class.

Amoxicillin by itself is susceptible to beta-lactamase attack, specifically penicillinases, rendering it ineffective.

Clavulanate provides no additional antibiotic coverage, but does protect the **amoxicillin** from beta-lactamases. Penicillin class antibiotics inhibit cell wall synthesis.

5. Provide the mechanism of action (M-O-A) of a tetracycline medication, two examples of tetracyclines, and two therapeutic uses.

Tetracyclines inhibit protein synthesis and include doxycycline and **minocycline**. Three uses for **tetracyclines** include periodontal disease, acne, and as part of a multiple antibiotic regimen in peptic ulcer disease.

6. Identify how sulfamethoxazole and trimethoprim work synergistically to kill bacteria and identify the life-threatening side effect sulfa medications can rarely cause.

Sulfamethoxazole and **trimethoprim** inhibit a bacteria's ability to produce folic acid at two different parts of the folic acid production cycle. Because humans can ingest folic acid and do not have to make their own, they are relatively unaffected. The life-threatening side effect associated with sulfa drugs is Stevens - Johnson syndrome (S-J-S), a hypersensitivity reaction involving the skin. Sulfa drugs can cause a tendency to burn easily when exposed to sunlight.

7. Provide a summary of the difference between gram-negative and gram-positive organisms and the effectiveness of two different cephalosporin generations against them.

Gram-positive and gram-negative organisms have different outer layers and as such, stain differently when processed in a lab. A gram-negative organism does not take up the Gram stain while a gram-positive organism does. Gram-positive organisms have a plasma membrane and thick peptidoglycan layer that retains the stain. In contrast, gram-negative organisms have a plasma membrane, a thin peptidoglycan layer, an outer polysaccharide, and a protein layer that does not take up the stain.

First-generation medications, such as **cephalexin**, work well against gram-positive organisms, but not gram-negative. **Cefepime**, a fourth-generation cephalosporin, works well against both. First-generation cephalosporins are narrow spectrum and fourth-generation cephalosporins broad spectrum with their additional coverage. Both produce bactericidal cell wall destruction.

8. Identify two medications, one bactericidal and one bacteriostatic, detailing how each medication works to create this effect based on their mechanism of action.

Bactericidal medications cause bacterial death, whereas bacteriostatic medications interfere with bacteria's reproductive process without killing the bacteria outright.

The bactericidal **penicillins**, **cephalosporins**, and **vancomycin** inhibit cell wall synthesis. Bacteriostatic **macrolides**, **tetracyclines**, and **clindamycin** affect ribosomes and replication.

9. Provide three unique types of antiviral medications and the pathophysiologic state each would treat.

Three antiviral types include those for human immunodeficiency virus (H-I-V), influenza, and herpes zoster. **Enfuvirtide** (brand, **Fuzeon**) comprises part of a multi-drug regimen to combat HIV. Influenza, if caught within the first 48 hours, yields to **zanamivir** or **oseltamivir**. Herpes zoster outbreaks respond to **acyclovir** and **valacyclovir**.

CHAPTER 5 - NEURO

1. Review the importance of sodium balance and lithium as it relates to their position on the periodic table.

Sodium (N-a) and lithium (L-i) align on the left side of the Periodic Table of Elements under group one, the alkali metals. Because they both work as cations with a plus one (positive charge), the body treats them similarly.

2. Review why you would use an ester or an amide as it relates to local parenteral anesthesia.

Two classes of local anesthetic include the esters and the amides. Esters and amides are functional groups from organic chemistry and provide a useful way to classify local anesthetics. Esters, when metabolized, end up as a para-aminobenzoic acid (PABA) metabolite. This is probably the source of many allergic reactions. If a patient has an allergy to PABA, they may also have issues with methylparaben, a preservative metabolized to PABA. As such, we generally limit esters for topical use. We prefer the amide **lidocaine** without that PABA issue.

3. Identify three classes of antidepressants. Provide an example of a drug in each class and a basis for the nomenclature, how it's named. Rank them from safest to least safe.

Three antidepressant classes include the selective serotonin reuptake inhibitors (SSRIs), the tricyclic antidepressants (TCAs), and monoamine oxidase inhibitors (MAOIs).

We name the SSRIs, such as **paroxetine** or **citalopram**, after the *single* neurotransmitter serotonin. The serotonin-norepinephrine reuptake inhibitors (SNRIs) like **venlafaxine** take their class name from *two* neurotransmitters. We name the tricyclic antidepressants, such as **amitriptyline**, after the shape of the molecule. A tricyclic molecule contains *three* rings. We name the MAOIs, such as **isocarboxazid**, after the enzyme they affect. Without regard to dosages, it's generally accepted that SSRIs are the safest antidepressants, followed by the TCAs and MAOIs.

4. Identify the advantages of a benzodiazepine over a barbiturate related to side effects. Provide an example of each medication giving generic and brand names.

The benzodiazepines, such as **alprazolam** (brand, **Xanax**) or **lorazepam** (brand, **Ativan**), provide anxiety relief more safely than a barbiturate such as **phenobarbital** (brand, **Luminal**) that can easily lead to a lethal overdose. Both are equally effective.

5. Identify four medications that can induce sleep or drowsiness from four unique medication classes.

Benzodiazepines such as **temazepam** and **diphenhydramine**, a first-generation antihistamine, cause drowsiness. **Barbiturates**, like phenobarbital, though dangerous, induce sleep. **Non-benzodiazepine** sedative-hypnotics like **eszopiclone** also help a patient rest.

6. Identify a common over-the-counter sleep aid and contrast that to a specific sedative-hypnotic as it relates to DEA considerations.

Over-the-counter sleep aids such as **acetaminophen with diphenhydramine** (brand, **Tylenol P-M**) and **diphenhydramine** (brand, **Benadryl**) are unscheduled. They do not carry addictive potential. However, the Drug Enforcement Agency (DEA) classifies sedative hypnotics, such as **zolpidem**, as schedule IV. This is a concern for patients with a history of narcotic abuse.

7. What generation holds high and low potency antipsychotics? Provide an example of a high and low potency antipsychotic. What are major differences between high and low potency antipsychotics' side effects and therapeutic effects?

The first generation distinguishes between high and low potency antipsychotics. For example, **chlorpromazine** (brand, **Thorazine**) is a low potency antipsychotic and **haloperidol** (brand, **Haldol**) is a high potency antipsychotic. That potency simply means the equivalent milligram strength for **haloperidol** will be smaller than **chlorpromazine**. An important distinction is that low potency medications tend to cause more sedation and less incidence of movement disorders (extrapyramidal symptoms (E-P-S)). High potency medications tend to cause less sedation and more E-P-S.

8. What are differences between typical and atypical antipsychotics as they relate to side effects and therapeutic effects? Provide an example of each.

Speaking broadly, the first-generation or typical antipsychotics like **haloperidol** have more movement disorders like the extrapyramidal symptoms and work only on positive symptoms of psychosis such as hallucinations. Second-generation (atypical) **antipsychotics** like **risperidone** help patients with both positive and negative symptoms. However, second-generation antipsychotics are more likely to cause diabetes, hyperlipidemia, and weight gain.

9. Contrast the treatment of a simple headache versus migraine, providing medication examples for the treatment of these two conditions. Discuss one medication used for migraine prophylaxis.

The over-the-counter analgesic **acetaminophen** or a nonsteroidal anti-inflammatory drug (NSAID), such as **ibuprofen** or **naproxen** relieves a simple acute headache. While **Excedrin Migraine** contains **acetaminophen, aspirin**, and **caffeine**, a very severe migraine might not respond to this over-the-counter remedy. A serotonin receptor agonist such as **sumatriptan** relieves an acute severe migranous attack. A prophylactic migraine medication might include a beta-blocker such as **propranolol**, a tricyclic antidepressant like **amitriptyline**, or an antiepileptic such as **divalproex**.

Chapter 6 - Cardio

1. Contrast the use of a beta1 adrenergic agonist and a beta1 adrenergic antagonist in two unique pathophysiologic states.

It's important to know the equation for cardiac output before tackling this question. Cardiac output equals stroke volume times heart rate.

A $beta_1$-agonist treats congestive heart failure or cardiogenic shock because the increase in heart rate should increase the cardiac output. What you do to the right side of the equation will influence the left side of the mathematical equation.

A selective $beta_1$-antagonist, like **metoprolol**, reduces hypertension and chronic increases in blood pressure. It might seem counterintuitive to use a beta-blocker in heart failure, because if you reduce heart rate on the right side of the equation, cardiac output should also go down. However, if the heart beats so fast it cannot fill, reducing the speed of the heart allows for more filling time and thereby improve stroke volume and cardiac output.

2. Provide a therapeutic indication (it does not have to be cardiac) for a Beta-1 antagonist and a Beta-2 agonist. Include two examples of Beta-1 antagonists and two examples of Beta-2 agonists. These can be combination medications.

A beta-1 selective antagonist like **metoprolol succinate** reduces hypertension. Another beta-blocker, **propranolol** prevents migraines.

A beta-2 agonist bronchodilates. Two examples include the rescue inhaler **albuterol** and the combination medication **fluticasone with salmeterol**. The salmeterol component is a long acting beta-2 agonist.

3. What do R-A-A-S, ACE, A-C-E-I and ARB stand for? Identify an A-C-E-I and an ARB.

The RAAS stands for the renin-angiotensin-aldosterone system. ACE is an acronym that stands for angiotensin converting enzyme (you pronounce the whole word). ACEI is an initialism (you pronounce each letter individually) that stands for angiotensin converting enzyme inhibitor. ARB is an acronym that stands for angiotensin II receptor blocker.
RAAS: renin-angiotensin-aldosterone system
ACE: angiotensin converting enzyme
ACEI: angiotensin converting enzyme inhibitor
ARB: angiotensin II receptor blocker

Lisinopril ends with the "pril, p-r-i-l" stem and is an ACE inhibitor. It prevents angiotensin-converting enzyme from converting angiotensin I to angiotensin II (the potent vasoconstrictor). **Valsartan** ends with the "sartan, s-a-r-t-a-n" stem, identifying it as an ARB. This medication blocks the receptors that angiotensin II binds to that produce vasoconstriction.

4. Identify a dihydropyridine and non-dihydropyridine C-C-B. Provide a rationale for the use of one over another in the treatment of dysrhythmias and uterine contractions.

A non-dihydropyridine calcium channel blocker might include **diltiazem** or **verapamil**. These CCBs affect the heart directly and are potent anti-dysrhythmics. However, these medications would not be used to prevent uterine contractions in a pregnant woman because of their cardiosuppressive effects on the mother's and fetus' hearts. Dihydropyridines like **nifedipine** reduce uterine contractions through calcium blockade. It's safer because its mechanism of action comes from vasodilation rather than cardiosuppression. This makes dihydropyridines inappropriate for dysrhythmias.

5. What do DCT and PCT stand for? Provide examples of a medication that works at each and outline potential adverse effects.

The distal convoluted tubule (D-C-T) and proximal convoluted tubule (P-C-T) are part of the nephron in the kidney. A diuretic that works at the PCT would include **mannitol**, for acutely raised intracranial pressure. Adverse effects could include dehydration and hypotension. **Hydrochlorothiazide** works at the distal convoluted tubule and causes hypokalemia. This is why manufacturers pair **hydrochlorothiazide** with **triamterene**, a potassium-sparing diuretic that counters the hypokalemic effect.

6. Identify the relative effects of HCTZ, furosemide, and spironolactone on water loss and the movement of potassium.

As a diuretic's site of action extends further from the glomerulus, less diuresis (water loss) is expected. In order of proximity to the glomerulus, **furosemide** works at the

Memorizing Pharmacology

Loop of Henle, **hydrochlorothiazide** at the distal convoluted tubule (DCT), and **spironolactone** at the collecting duct.

As such, **furosemide** provides the most diuresis and causes the greatest loss of potassium. **Hydrochlorothiazide** produces less diuresis than a loop diuretic, but more than a potassium-sparing diuretic and has the potential to cause hypokalemia. The potassium-sparing diuretic **spironolactone** causes only moderate diuresis, but may lead to hyperkalemia, not hypokalemia.

7. Identify three pathophysiologic states that benefit from diuresis. They don't have to be cardiac. Provide three diuretic medications used for those states.

Treat uncomplicated hypertension with **hydrochlorothiazide** and a potassium-sparing diuretic like **triamterene** or the combination medication **hydrochlorothiazide with triamterene**. The edema (swelling due to excess water) secondary to congestive heart failure responds to **furosemide**, a loop diuretic. **Mannitol** reduces acutely raised intracranial pressure.

8. Provide the advantages and disadvantages of warfarin, heparin, and enoxaparin.

Warfarin has the advantage of oral administration but requires monitoring of the international normalized ratio (I-N-R). The INR measures blood coagulation and helps a health care professional make dosage adjustment decisions.

Patients can take the injectable **enoxaparin** home. **Heparin** requires more frequent dosing, injections, monitoring and hospitalization, but is relatively inexpensive.

9. Include an HMG-CoA medication adversely affected by grapefruit juice and the potential effects of this interaction.

Grapefruit juice affects the metabolism in some drugs, such as the HMG-CoA reductase inhibitor **atorvastatin**. This effect is less so or nonexistent with **rosuvastatin**. Grapefruit juice also affects the calcium channel blockers **nifedipine** and **verapamil**. As with the HMG-CoAs, within a class of medications, one drug might interact, while another does not.

10. Outline two potential pathophysiologic complications of HMG-CoA reductase inhibitors. Identify two unique HMG-CoAs.

Muscle soreness or pain appears in about 1 in 10 patients, with the rarer side effects including myopathy (more severe muscle pain) and rhabdomyolysis (a muscle tissue breakdown). Two HMG-CoAs include **rosuvastatin** and **atorvastatin**.

CHAPTER 7 - ENDOCRINE / MISC.

1. Provide two advantages of the use of glucophage (Metformin) over other antidiabetic medications.

Glucophage generally has a low incidence of hypoglycemia and can induce a small amount of weight loss during treatment. In addition, it reduces L-D-L (bad cholesterol) and slightly raises H-D-L (good cholesterol) levels.

2. Provide three signs and or symptoms of hypo- and hyperthyroidism and a medication used for each condition.

Hypothyroid patients present with chronic fatigue, constipation, weight gain, and bradycardia. Generally, supplementation is effective with **levothyroxine**. Low iodine may also be a cause. A hyperthyroid patient might present with the opposite symptoms: nervousness, tachycardia, anxiety, and insomnia. **Propylthiouracil** is a preferred treatment.

3. Identify two insulin types, their time course of action, and one special consideration.

Insulin glargine works for the entire day and improves the blood sugar profile of a diabetic. **Humulin R**, a faster acting insulin, works with an insulin pump.

4. Identify six anticholinergic effects and the opposing six cholinergic effects. Identify one anticholinergic medication and one medication that would cause cholinergic effects and their therapeutic uses.

Anticholinergic effects fall under the broad category of dryness, i.e., <u>a</u>nhidrosis, <u>b</u>lurry vision (secondary to dry eyes), <u>d</u>ry mouth, <u>u</u>rinary retention, <u>c</u>onstipation, and <u>t</u>achycardia. (Remember the **abduct** water mnemonic)

Cholinergic effects include sweating, lacrimation (watery eyes), hypersalivation, urinary incontinence, diarrhea, and bradycardia.

An anticholinergic medication includes **tolterodine** (brand, **Detrol**) for urinary incontinence. The brand name refers to "cont<u>rol</u>ling the <u>det</u>rusor" muscle over activity. **Bethanechol** (brand, **Urecholine**), a cholinergic, relieves urinary retention. The brand name might allude to <u>ur</u>inary retention and the <u>cholin</u>ergic receptors affected.

5. Identify two physiologic systems that alpha-1 blockers affect. Provide two examples of alpha-1 blocking medications. Note how their brand names allude to their therapeutic effect.

Alpha-blockers work for benign prostatic hyperplasia (B-P-H) and hypertension. **Tamsulosin** (brand, **Flomax**) provides "<u>max</u>imum <u>flo</u>w of urine" by relaxing the smooth muscle in the prostate. **Doxazosin** (brand, **Cardura**) helps improve hypertension and affects the "card" iovascular system.

6. Provide a drug interaction that might occur with sildenafil (Viagra) and a vasodilator.

There is a potential for a dangerous sudden decrease in the systemic blood pressure with **nitroglycerin (brand, Nitrostat)**.

7. Provide two examples of birth control medications that fall outside of the traditional 21 active pills, 7 inactive pills.

Ortho Evra (norelgestromin with ethinyl estradiol) is a patch that's put on the upper outer arm, abdomen, or buttocks for 21 days, removed for 7 days, and then a new patch is applied.

Nuvaring (etonogestrel with ethinyl estradiol) is a contraceptive ring that's inserted vaginally for 21 days. The patient has seven drug free days and then inserts another ring.

EPILOGUE

WHY MEMORIZATION MATTERS

The tone of this book is one of relaxation, a step-wise and logical progression from easy to hard material with a bit of narrative and humor.

In the print book, I didn't share what prompted me to become so passionate about extending this information – a terrible test of my competence as a father and health professional and a clear need for memorization.

My intention was to create an audiobook from the print book that I would read. I couldn't include this story because I can't read it myself without breaking down completely. Writing this essay provided a means for me to deal with the emotional trauma of that terrible night.

I now have a narrator and would like to share this story in the eBook and audiobook, originally published in an outstanding literary journal – *Intima: A Journal of Narrative Medicine* in the spring 2014 issue.

THERE WILL BE NO PROBLEMS

Baby C has had it tough from the time she was inside the womb, a quiet girl flipping in the corner, kicking from time to time, but mostly playing by herself until her sister, Baby A, tried to break out at 19 weeks. In a sweat-soaked hospital bed, I held my wife's hair as she vomited into a Cinderella pink bucket. Hours before, she volunteered to have magnesium sulfate infused into her vein to stop her contracting uterus, the shell loosely holding our unborn triplet daughters. My wife's shivers turned violent as the doctor doubled the dose trying to delay the preterm contractions in hopes that a stitch could be inserted, shutting her incompetent cervix. Should we have allowed the doctor to stab a javelin of potassium chloride into a Doppler pounding fetal heart, reducing and aborting one daughter that the others might have a chance to live? Did we sentence our daughters to die at 19 weeks because we kept all three?

The news came that morning from our perinatologist. The contractions slowed enough that he was able to push Baby A back in, tie the cervix with a cerclage procedure, and put mom on hospital bed rest. Days turned into weeks and into months until week 27 and 2 days when the dam broke. I posted "babies are coming" on our Facebook pages and the streams of electronic congratulations came in. A painful juxtaposition to what we knew about babies that are born before 28 weeks. A well-intentioned neonatologist told us "you must make it to week 28."

We learned each week closer to 28 lessens the risk of retinopathy of prematurity (R-O-P), necrotizing enterocolitis (NEC, N-E-C), patent ductus arteriosus (P-D-A), respiratory distress syndrome (R-D-S), intraventricular hemorrhage (I-V-H), or of finding the children cannot survive in this world and die. We toured the NICU and after seeing two children the size of a dollar bill, I asked that we return to her room. My wife's only time off hospital bed rest is a 15- minute wheelchair ride every 48 hours. She turns it down.

The anesthesiologist, a likeable Duke graduate, ended up playing the role of husband assuaging my wife while I called the play-by-play to her by looking over the sheet that served as a screen. A medical resident had his leg up on the table to leverage the retractor that opened my wife's antiseptic covered yellow belly. I had asked if he could deliver the babies vaginally and the doctor assured me he could, if I could get them to agree to all come out headfirst.

EPILOGUE

It was not the scalpel making the bikini cut, the Pfannenstiel incision or the creeping blood, or even the exposed peritoneal cavity that worried us. Silence did.

We held hands and welcomed the three screams that came from the girls not much larger than a soda can. It would be their last cry for weeks. As the perinatologist closed my wife's abdominal wall with sutures, I went up to their room. Seventeen people were in their triplet suite, a corner office with three parallel Giraffe incubators lit up like tanning beds, bathing the girls in U-V rays that would reduce their bilirubin, a danger to their brains. A machine would breathe for them. A machine warmed them. A machine fed them.

I sign a release form for their upcoming blood transfusions. I'm told to come back in half an hour. I return in 10 minutes after talking to the family and get kicked out. I find my wife loosely thumbing the green button of a morphine pump. A nurse tries to give me a congratulatory hug. I retreat, crashing into a wall to avoid being touched.

Baby A, long and lithe, the giant at two pounds, looked to be the healthiest. Baby B, a scrapper who fought every needle and prod, matched her sister Baby C at one pound six ounces. Baby C lay swaddled by the window, a place for the parents to look outside and away. Baby C kept having apneic spells where she desatted, quietly trying to slip away in the shadow of the red alarm. The cardiologist found a hole in a blood vessel above her heart, patent ductus arteriosus, and if medicine did not close it, he would have to operate. A second round of indomethacin, an anti-inflammatory, finally relayed the message to the patent ductus that the baby was no longer in the womb, and could close.

Baby C still struggled from time to time, an idiopathic infection that came and went and reflux that burned her esophagus. She had trouble catching up with her sisters who needed less oxygen support. From time to time Baby C would vomit her milk, usually half an hour after a feeding.

This last emetic episode came as she was alone in a room that her sisters had already vacated for home. Weeks later, the hospital staff felt they could trust us to take her home, that we knew what we were doing, and that she would be fine in our care – a third child with a third apnea monitor.

The apnea monitor is a box a little bigger than a hardcover book that fires a piercing scream meant to both wake the child and the parent when it stops detecting motion. Imagine putting your ear to an eighteen-wheeler's horn as it blows. It's easy to get comfortable with the false alarms; it's easy to want to turn them off.

Baby C lay quietly next to me on the floor of the nursery as we watched Iowa State University's unranked football team take a 21-to-0 lead on Texas Tech on a small television screen. The camera panned to a small contingent of fans excited they made the trip to Lubbock. Nothing could go wrong for the Cyclones.

In frank juxtaposition, Baby C's baby monitor fired, waking her sisters who started their own punishing screams in the cribs above. Baby C did not cry, but vomited a small amount of milk and began to turn purple. I called to her, asking her to breathe for me, screamed for my wife and began CPR. Well, I tried CPR.

When would we ever actually use CPR? The practice baby in class felt like a toy, a hard plastic shell whose chest pushed down neatly when I did the 100 compressions. My stomach knotted from lunch that day, so I excused myself a couple of times, missing some content, like trying to read a book, watch television and surf the net at the same time.

On our nursery floor, Baby C felt limp, like a stuffed animal, an infant turning crimson in her nightclothes. I ripped the sleeper to find the tiny xiphoid process and started to press down with my two fingers wiping the sting from my eyes. I screamed for my wife. No answer. I put a breath in her tiny mouth, but it hit a wall to spurt some vomit from her nose into her oxygen cannula. I screamed again for my wife who I later found was in the master bedroom brushing her teeth with an electric toothbrush, and could not hear.

"What do you want me to do?" she asked.

"Call 9-1-1," I said.

My father saved my life when I spiked a fever so high he had to submerge me kicking and screaming into a tub of water filled with ice. My father passed his test.

Together, we worked. Compressions and breaths until one of us remembered: An infant should have their nose and mouth covered when putting in a breath. Newton's third law, an equal and opposite reaction, pushed the curdled milk against the blocked pyloric valve, bouncing it back into her stomach and out her esophagus. A pool of vomit and tiny violin strings of blood covered the mat. I followed the little green hose to her oxygen concentrator running it up to five liters.

The haze of the policeman, the paramedics, my wife, my future life without my daughter, my life with my daughter, an infant in the back of the ambulance, the minutes before the one-hour surgery to correct, the pediatric surgeon's enormous hands, the surgeon's baritone words.

"There will be no problems," he said. "But . . . ,"

"There will be no problems."

GENERIC AND BRAND NAME INDEX

aba<u>tacept</u>, 22, 23, 28, 153
Abreva, 47
acetaminophen, xxxiv, xlvi, liv, 15, 16, 18, 19, 28, 77, 99, 153, 156
acetaminophen PM, 99, 156
a<u>cyclovir</u>, 63, 71, 155
Advair, 36, 37
Advil, xxviii, 15
Afrin, 32
albu<u>terol</u>, xxxiii, 37, 38, 43, 154
albu<u>terol</u> / ipra<u>tropium</u>, 38
Aldactone, 102, 105
alen<u>dronate</u>, 23, 28, 153
Aleve, 15
alfuzosin, 140, 144, 158
allopurinol, 25, 28, 153
alp<u>razolam</u>, 85, 86, 99, 156
Ambien, 77, 78
ami<u>kacin</u>, 56, 71, 155
Amikin, 56
ami<u>triptyline</u>, 80, 83, 99, 156
amlo<u>dipine</u>, 114, 115, 125, 157
amoxi<u>cillin</u>, xix, 48, 49, 50, 71, 155
amoxi<u>cillin</u> / clavulanate, 50
Amoxil, xix, 49, 50
amphotericin B, 61, 71, 155
Anbesol, 74
AndroGel, 135, 136

Antivert, 74
Aricept, 95
aspirin, xlvi, lvi, lvii, 14, 16, 28, 102, 125, 153, 157
aten<u>olol</u>, 110, 125, 157
Ativan, 86
atom<u>oxetine</u>, 87, 88, 99, 156
ator<u>vastatin</u>, 117, 125, 157
Atripla, 65, 66, 70, 149
AtroPen, 121
<u>atropine</u>, 120, 121, 125, 157
Augmentin, 50
Avodart, 140
azi<u>thromycin</u>, 54, 71, 155
Benadryl, 30
Benicar, 113, 114
benzo<u>caine</u>, liv, 74, 99, 156
bethanechol, 138, 144, 158
Biaxin, xix, 54
bismuth sub<u>sa</u>licylate, 6
Boniva, 24
budesonide, 36, 43, 154
budesonide / form<u>oterol</u>, 36
butenafine, 46, 71, 155
Calan, 6, 85, 115
carbamaze<u>pine</u>, 92, 99, 156
Cardizem, 115
Cardura, 108
carve<u>dilol</u>, 108, 111, 125, 157
Catapres, 109
<u>cef</u>epime, 51, 71, 155

189

ce<u>f</u>triaxone, 51, 71, 155
Celebrex, 16
cele<u>coxib</u>, 16, 28, 153
Celexa, 80, 81
ce<u>ph</u>alexin, 50, 71, 155
cetirizine, xlviii, 31, 43, 154
Chantix, 84
Cheratussin AC, 34
chlorpromazine, 90, 91, 99, 156
Cialis, xxxii, 139
Cipro, 58
cipro<u>floxacin</u>, xxxiv, 58, 71, 155
citalopram, 80, 99, 156
clari<u>thromycin</u>, xix, 54, 71, 155
Claritin, 30, 31
Claritin-D, 30, 31
Cleocin, 55
clinda<u>mycin</u>, 55, 71, 155
clon<u>azepam</u>, 86, 99, 156
clonidine, 109
clopido<u>grel</u>, 120, 125, 157
Colace, 7
Concerta, 87, 88, 109
Coreg, 111
Coumadin, 119
Cozaar, 113
Crestor, 117
cyclobenzaprine, 24, 28, 153
Cymbalta, 81, 82
dabi<u>gatran</u>, 118, 119, 120, 125, 157
daru<u>navir</u>, 66, 155
Deltasone, 36
Depakote, 93
desvenla<u>faxine</u>, 82
Detrol, 138

dexmethylphenidate, xxxiii, 87, 99, 156
DiaBeta, 132
di<u>azepam</u>, 24, 28, 153
Diflucan, 61
digoxin, 120, 121, 125, 157
Dilantin, 93
diltiazem, 115, 125, 157
Diovan, 114
diphenhydramine, xxxiv, xli, xlviii, 29, 30, 43, 77, 154
Ditropan, 138
divalproex, 93, 99, 156
docosanol, lii, 47, 71, 155
docusate sodium, xli, 7, 12, 152
done<u>pezil</u>, 95, 99, 156
Doryx, 53
dox<u>azosin</u>, 107, 108, 125, 157
doxy<u>cycline</u>, 53, 71, 155
Dramamine, 74
duloxetine, 82, 99, 156
DuoNeb, 38, 42, 147
Duragesic, 18
duta<u>steride</u>, 140, 144, 158
Dyazide, 104
Ecotrin, 14, 15, 102
efa<u>virenz</u>, 65, 71, 155
Effexor, 82
Elavil, 83
Eldepryl, 94, 95
ele<u>triptan</u>, 21, 28, 153
emtri<u>citabine</u> / tenofo<u>vir</u>, 65
E-Mycin, 53, 54
enalapril, 113, 125, 157
Enbrel, 23
enfu<u>virt</u>ide, 65, 71, 155
enoxa<u>parin</u>, 118, 125, 157
epinephrine, 40, 43, 106, 154

GENERIC AND BRAND NAME INDEX

EpiPen, 40
erythromycin, 54, 71, 155
escitalopram, 80, 99, 156
esomeprazole, 4, 12, 152
estradiol, 144, 158
eszopiclone, 76, 77, 99, 156
etanercept, 23, 28, 153
ethambutol, 60, 71, 155
Excedrin Migraine, 14, 16, 20, 27, 146
famotidine, 3, 12, 152
febuxostat, 25, 28, 153
fenofibrate, 117, 125, 157
fentanyl, xxxiii, 18, 28, 153
finasteride, 140, 141, 144, 158
Flagyl, 59
Flexeril, 24
Flomax, 140
Flonase, 37
Flovent Diskus, 37
Flovent HFA, 37
fluconazole, 61, 71, 155
Flumist, 46, 47
fluoxetine, 81, 99, 156
fluticasone, 37, 43, 154
Fluzone, 46, 47
Focalin, 87
Fosamax, 23
Fungizone, xxvi, 61
furosemide, xxxiv, 103, 125, 157
Fuzeon, 65, 70, 149
gabapentin, 93, 99, 156
Garamycin, 56
gentamicin, 56, 71, 155
glipizide, 132, 144, 158
GlucaGen, 132, 133
glucagon, 132, 144, 158
Glucophage, 130, 131

Glucotrol, 132
glyburide, 132, 144, 158
guaifenesin / codeine, 34
guaifenesin/DM, 43, 154
Haldol, 91
haloperidol, 90, 91, 99, 156
heparin, 119, 125, 157
Humalog, 128, 130, 133, 142, 151
Humulin N, 129, 131
Humulin R, 128, 130
hydrochlorothiazide, xxxiii, 104, 125, 157
ibandronate, 24, 28, 153
ibuprofen, xxviii, xlvi, 15, 28, 153
Imitrex, 21
Imodium, 7
Inderal, 109
infliximab, 8, 9, 12, 152
influenza vaccine, lii, 71, 155
INH, 60
insulin glargine, 133, 144, 158
insulin lispro, 133, 144, 158
Isentress, 66, 70, 149
isocarboxazid, 83, 99, 156
Januvia, 131
Kadian, 17, 18
Keflex, 50, 51
Klonopin, 86
Lamictal, 85
Lanoxin, 120, 121
Lantus, 131, 133, 134, 143, 151
Lasix, 103
Levaquin, 58
levetiracetam, xxxiv
levodopa / carbidopa, 94
levofloxacin, 58, 71, 155

191

levonorgestrel, lvii, 128, 129, 144, 158
levothyroxine, xxxiii, 134, 144, 158
Lexapro, 80, 81
lidocaine, liv, 74, 99, 156
linezolid, 53, 55, 71, 155
Lipitor, 117
lisinopril, 113, 125, 157
lithium, 89, 99, 156
Lithobid, 89
Loestrin 24 Fe, 135, 136, 143, 144, 151, 158
loperamide, xxxiv, xli, xliv, 7, 12, 152
Lopressor, xxix, xxx, 110
loratadine, xlviii, 31, 43, 154
loratadine-d, 43, 154
loratadine-D, xlviii
lorazepam, 86, 99, 156
losartan, 113, 125, 157
Lotrimin Ultra, 46
Lovaza, 101
Lovenox, 118
Lunesta, 76, 77
Lyrica, 93
Magnesium Hydroxide, 2
mannitol, 103, 125, 157
maraviroc, 65, 71, 155
Marplan, 80, 83
Maxipime, 51
meclizine, liv, 74, 98, 99, 149, 156
Medrol, 35
meloxicam, 14, 16, 28, 153
memantine, 95, 99, 156
metformin, 131, 144, 158
methotrexate, 22, 28, 153

methylphenidate, xxxiii, 88, 99, 156
methylprednisolone, 35, 36, 43, 154
metoprolol succinate, 110, 111
metoprolol tartrate, 110
metronidazole, 57, 59, 71, 155
Microzide, 104
midazolam, 86, 99, 156
Milk of Magnesia, 2
Minocin, 53
minocycline, 53, 71, 155
MiraLax, 7
Mobic, 16
montelukast, 35, 39, 43, 154
morphine, 17, 28, 153
Motrin, xxviii, xxix, 15
MS Contin, 17, 18
Mucinex DM, 33
Myambutol, 60
Mycostatin, 61, 62
naloxone, 20, 28, 153
Namenda, 95
naproxen, xlvi, 15, 28, 153
Narcan, 20, 27, 146
Nasacort Allergy 24HR, 32, 33
neomycin / polymyxin B / bacitracin, 46
Neosporin, 46
NeoSynephrine, 32
Neurontin, 93
Nexium, 4, 5
niacin, lvi, 101, 125, 157
Niaspan ER, 101
nifedipine, 116, 125, 157
nitroglycerin, 116, 124, 125, 150, 157
Nitrostat, 116, 124, 150
Norvasc, 115, 116

GENERIC AND BRAND NAME INDEX

NPH insulin, 144, 158
NuvaRing, 135, 137, 143, 144, 151, 158
nystatin, 61, 71, 155
olmesartan, 113, 125, 157
omalizumab, 35, 39, 43, 154
omega-3, lvi, 101, 124, 125, 150, 157
omeprazole, xix, xxxiv, 5, 12, 152
ondansetron, 8, 12, 152
Orencia, 22
OrthoEvra, 135, 136, 143, 144, 151, 158
oseltamivir, 62, 71, 155
Osmitrol, 103
oxybutynin, 138, 144, 158
oxycodone, 18, 19, 28, 153
oxymetazoline, 32, 43, 154
palivizumab, 64, 71, 155
paroxetine, 82, 99, 156
Paxil, xxx, 82
Paxil CR, 82
penicillin, 50
Pepcid, xxxviii, 3
Pepto-Bismol, xxvii, 2, 6
Percocet, 19
Phenergan, 8
phenylephrine, 32, 43, 154
phenytoin, 93, 99, 156
Plan B One-Step, 129, 142, 151
Plavix, 120
polyethylene glycol, 12, 152
potassium chloride, 105
Pradaxa, 119
prednisone, 36, 43, 154
pregabalin, 93, 99, 156
Prezista, 66, 70, 149
Prilosec, xix, xxxvi, 5

Pristiq, 82
ProAir HFA, 37, 38
Procardia, 116
promethazine, 8, 12, 152
Propecia, 140, 141
propranolol, 109, 157
propylthiouracil, 134, 144, 158
Proscar, 140
Prozac, 81, 82
pseudoephedrine, xli, xlviii, 31, 43, 154
PTU, 134, 143, 151
pyrazinamide, 60, 71, 155
PZA, 60
quetiapine, 92, 99, 156
ramelteon, 78, 99, 156
ranitidine, xli, 4, 12, 152
regular insulin, 144, 158
Relenza, xxxii, 63
Relpax, 21
Remicade, 8, 9, 64
Rheumatrex, 22
Rifadin, 60
rifampin, 60, 71, 155
Risperdal, 91
risperidone, 91, 99, 156
Robitussin DM, 33
Rocephin, 51
rosuvastatin, 117, 125, 157
Rozerem, 77, 78
salmeterol / fluticasone, 36
Sarafem, 81, 82
scopolamine, 96, 99, 156
selegiline, 94, 99, 156
Selzentry, 65, 70, 149
Seroquel, 92
sertraline, 81, 99, 156
sildenafil, 139, 144, 158
Sinemet, 94

193

Singulair, 39
sitagliptin, 131
Solarcaine, 74
solifenacin, 138, 144, 158
Spiriva, 38
spironolactone, xxxiv, 105, 125, 157
Strattera, 81, 88
Sublimaze, 18
Sudafed, 31, 32
sulfamethoxazole / trimethoprim, 57
sumatriptan, 21, 28, 153
Symbicort, 36
Synagis, 64
Synthroid, 134
tadalafil, 139, 144, 158
Tamiflu, xxxii, 47, 62, 63
tamsulosin, 140, 144, 158
Tegretol, 92, 93
Tenormin, 110
testosterone, 135, 143, 144, 151, 158
Thorazine, 90
tiotropium, 35, 38, 43, 154
tolterodine, 138, 144, 158
Toujeo, 131, 133
tramadol, xxxiv, 19, 28, 153
Transderm-Scop, 96
triamcinolone, xli, 33, 43, 154
Tricor, 117
Tri-Sprintec, 135, 136, 143, 144, 151, 158
Tums, xxvii, 2
Tylenol, xxvi, 15, 19, 75
Tylenol PM, 75, 77

Uloric, 25
Ultram, xxxi, 19, 20
Urecholine, 138, 139
Uroxatral, 140
valacyclovir, 63, 71, 155
Valium, xxxii, 24
valsartan, 114, 125, 157
Valtrex, xxix, 63
Vancocin, 52
vancomycin, 49, 52, 71, 155
varenicline, 84, 99, 156
varicella, 63
Varicella- Zoster, 63
Vasotec, 113
venlafaxine, 82, 99, 156
verapamil, 115, 125, 157
Versed, 86
VESIcare, 138
Viagra, xxxii, 139
Vicodin, 18, 19
warfarin, 119, 125, 157
Wellbutrin, 84
Xanax, 85, 86
Xolair, 39, 40, 64
Zantac, 4
Zestril, 113
Zithromax, 54
Zofran, 8
Zoloft, 81
zolpidem, 78, 99, 156
zoster, 63
Zovirax, 63
Zyban, 84
Zyloprim, 25
Zyrtec, 31
Zyvox, 55

CPSIA information can be obtained
at www.ICGtesting.com
Printed in the USA
BVHW072000020920
587789BV00002B/105